LOCKHEED CONSTELLATION

Curtis K. Stringfellow and Peter M. Bowers

Motorbooks International
Publishers & Wholesalers ®

First published in 1992 by Motorbooks International Publishers & Wholesalers, PO Box 2, 729 Prospect Avenue, Osceola, WI 54020 USA

The information in this book is true and complete to the best of our knowledge. All recommendations are made without any guarantee on the part of the author or Publisher, who also disclaim any liability incurred in connection with the use of this data or specific details

We recognize that some words, model names and designations, for example, mentioned herein are the property of the trademark holder. We use them for identification purposes only. This is not an official publication

Motorbooks International books are also available at discounts in bulk quantity for industrial or sales-promotional use. For details write to Special Sales Manager at the Publisher's address

Library of Congress Cataloging-in-Publication Data
Stringfellow, Curtis K.
    Lockheed Constellation / Curtis K. Stringfellow, Peter M. Bowers. p.   cm.
    ISBN 0-87938-379-8
    1. Constellation (Transport planes)   I. Bowers, Peter M.   II. Title.
TL686.L6S77   1992
387.7'3343—dc20   89-12809

Printed and bound in Hong Kong

On the front cover: President Dwight Eisenhower's C-121A *Columbine II,* the first Air Force One, was saved from the scrap heap and lovingly restored by Harry Oliver of Santa Fe, New Mexico. *Bob Shane*

On the frontispiece: C-69s on the outdoor modification line at the Lockheed Air Terminal at Burbank. *Lockheed*

On the title page: The Lockheed Model 749A-79 Constellation in this revealing photo represents the many Connies that have had several owners and different national registrations during their lifetimes. HK-652 was originally delivered to the Dutch airline KLM as Model 749-79-33 PH-TDH in November 1947, and was soon converted to a Model 749A. It was sold to Capital Airlines in the United States as N90608 in July 1952, and then went to British Overseas Airways Corporation (BOAC) as G-ANVA in May 1955. Avianca in Colombia bought it as HK-652 in April 1959. In 1962 it returned to the United States as N6687N and was converted to a freighter. It was finally sold to Peru, but was not used before being scrapped in 1969. *Gordon S. Williams*

On the back cover: Upper left, Panair do Brasil, a subsidiary of Pan Am, operated sixteen Model 049s. *Lockheed.* Upper right, a Model 049 on a Lockheed test flight before the aircraft was delivered to TWA on May 15, 1947. *Lockheed.* Lower left, a US Navy WV-2 AEW aircraft. *Lockheed.* Lower right, a Model 1049G Super Constellation in the livery of the Spanish airline Iberia. *Lockheed*

Contents

Acknowledgments

Much of the basic material for a book like this comes from a few standard sources such as the public relations offices of the manufacturer, the airlines, and the armed services. However, additional information and data that contribute greatly to the overall picture, such as personal experiences, postdelivery photos, and so on, come from private individuals. It is these people—many of whom we have never met personally and have loaned us such material without compensation—who we wish to acknowledge here in addition to the primary sources:

Harold Andrews
J. Roger Bentley
Ed Betts (TARPA Topics)
John Campbell
Liwa Chiu (Pan American)
Mrs. Gerald Fay
Clint Groves (ATP/Airliners America)
Dave Kellogg (Pacific Northern Airlines)
Gary Killian (FAA)
David W. Lucabaugh
April McKittrick (Lockheed)
David W. Menard
Betty Moore (Delta Air Lines)
Douglas D. Olson
Ed Peck
Jon Proctor
Walter J. Redmond
Boardman C. Reed
Victor Seagel (TWA)
Victor D. Seely
Eric Shulzinger (Lockheed)
Smithsonian Library Staff
Norman E. Taylor
Ray Wagner (San Diego Aerospace Museum)
John Wegg (*Airliners* Magazine)
John B. White III

Researching and writing a book also takes time away from one's normal family activities. We wish to express gratitude to our families for their patience—our wives Joan Stringfellow and Alice Bowers, and daughter Kristen Stringfellow.

Curtis K. Stringfellow and
Peter M. Bowers

Northwest Orient Airlines initially placed an order for six 1049Gs but later reduced it to four. These Connies were configured for seventy-four passengers and were flown exclusively from Seattle to Tokyo and Hawaii. Lockheed

Introduction

The plan for the Constellation was conceived in April 1939 by Transcontinental & Western Airlines (TWA, later to become Trans World Airlines), which was then under the financial control of Howard Hughes. Hughes wanted an airliner with 300 plus mph speed that was capable of making nonstop eastbound flights in eight hours or one-stop westbound flights. After making numerous visits to aircraft manufacturers about production of such an airplane, Hughes found that Lockheed was interested in submitting a proposal.

Initially, Lockheed proposed an airliner already under development, but TWA wasn't interested so a new design was drawn up. Lockheed had never built an airliner the size of the Constellation before, but this did not hinder its efforts. Under the expert guidance of designer C. L. "Kelly" Johnson, the prototype was completed by the end of 1942. The airplane was larger than most in production at that time, so several key components were among the first to be tried on an airliner. These included the use of hydraulically boosted controls, full pressurization, and high-lift flaps on the wings.

Because of subsequent publicity some believed that Hughes himself actually designed the Constellation, but in fact he only drew up the initial specifications while Lockheed did the designing. Following the outbreak of war the Constellation, or Connie as it is popularly called, became the C-69 for the United States Army Air Forces (USAAF). After being transferred to the USAAF for testing, the Constellation completed the trials in only thirty-three days—the shortest time of any large airplane up to that period. While completing the accelerated service trials, the C-69 had the distinction of carrying Orville Wright on what would turn out to be his last flight.

After the war ended, numerous airlines flocked to Lockheed to buy the Constellation. Unfortunately for many domestic airlines, and also for Lockheed, Hughes' contract stated that no Constellation could be purchased by a domestic east-west operator for a period of two years. While this sent many domestic airlines to Lockheed's competitors, it did not stop the overseas and foreign airlines from purchasing it.

From then on there was no looking back for the Connie as it set many speed and distance records. Newer models such as the 649 and 749 arrived on the scene soon after the war, providing improved accommodations and performance. Unfortunately for most of the airlines, a significant decrease in passengers occurred after the war, forcing many to either cut back on their orders for new airliners or cancel them altogether. At one point things were so bad that Lockheed even considered closing down the Constellation production line. Luckily, the US military was looking for a new transport and ordered new versions of the 749. This kept the production lines open and also helped to pave the way for such future developments as the Super Constellation.

The Constellation was developed initially for the airlines to provide increased seating capacity, but it was also eagerly sought by the military. The addition of the newly developed turbo-compound engines brought the Connie into its own with substantial increases in economy and performance. With the military proving the concept of the turbo-compounds, the airlines soon began to acquire this new technology and were finally able to offer true nonstop transcontinental services across the United States. Special freighter versions that catered to the cargo airlines were also developed.

The ultimate piston-engine airliner arrived with the Model 1649A Starliner. This airplane allowed the airlines to offer nonstop flights to just about anywhere in the world year-round. Unfortunately this model arrived too late, however, and was quickly rendered obsolete by the newer jets.

With the introduction of jet-propelled airliners, most of the airlines began to sell their older piston-engine airplanes. Most were sold to other airlines but some went to travel clubs or even to individuals. Many of the airplanes had varied and interesting careers before finally being declared unairworthy and set out to pasture. Some fine examples can still be seen today in all parts of the world, some still flying and some on static display. But this number is gradually dwindling and soon none will be left to enjoy. Most aviation buffs agree that the Constellation is one of the most beautiful propeller transport airplanes ever produced.

To most, the Constellation was the epitome of piston-engine airliners, yet to others it was just another airplane. Most pilots agree that it was a pleasure to fly but some say it was the best three-engine airliner in the world, at a time when engine failures were common. No matter what, everyone has an opinion about the Constellation that is uniquely his or her own. In this book we have tried to remain neutral about the airplane, showing all sides of it and even delving into some personal experiences. To the airlines that operated the Constellations we dedicate this book, for without their keen insight there wouldn't be any Connies to cherish today.

The Airlines of America, 1927-1941

An extensive network of airlines evolved in Europe right after World War I. After a brief period of using converted war-surplus bombers, dedicated single- and multi-engine transports evolved to meet increasing passenger demand.

Equivalent airlines did not emerge in the United States until 1927, after the government turned the last of its airmail routes over to private contractors. Many smaller firms had attempted to operate passenger-only point-to-point scheduled air service since the war, but all such efforts failed because they could not prosper on passenger revenue alone. The revenue from airmail, which helped ensure the success of the European airlines, was not available to the struggling US operators; the government was carrying the mail itself in its own airplanes.

The Awakening

Legislation to end governmental airmail practices was passed in 1925, and government routes were transferred to civil operators early in 1926. At first these contractors carried mail only in single-engine airplanes, similar to the ones used by the post office. Then in July 1927, Boeing Air Transport, a subsidiary of the Boeing Airplane Co. of Seattle, Washington, started mail and passenger service from San Francisco to Chicago in single-engine Boeing Model 40A biplanes that featured a two-passenger cabin ahead of the pilot's traditional open cockpit behind the wing. The later Model 40B-4 had a four-passenger cabin. Successful American passenger-and-mail operations date from Boeing's innovation.

Large and small airlines alike used a variety of single-engine cabin monoplanes, and Lockheed became outstanding in the single-engine field. As passenger traffic increased on the trunk routes multi-engine models, most notably the Fokker F-7 and F-10 trimotor monoplanes and the very similar Fords, were modified for passenger capacities of ten to fourteen. The use of single-engine airliners on the trunk lines was ruled out by legislation passed in 1934, but Lockheed had pulled out of that market by that time.

Boeing's Revolution

Early in 1933 Boeing revolutionized the airline business with its innovative Model 247, a ten-passenger low-wing, twin-engine, all-metal monoplane with retractable landing gear and cowled 550hp Pratt & Whitney R-1340 radial engines. These were the same proven engines already in use by Fokker and Ford.

While the new Boeing cruised some 40mph faster than the old trimotors on less power, and was a great step forward in technology, it did not go quite far enough. Its passenger capacity was too few for the size of airplane, and while new in many respects it retained such old concepts as not being equipped with wing flaps. The early models also retained fixed-pitch propellers.

Even before the 247 flew, its details became well known throughout the industry, and the Douglas Aircraft Co. of Santa Monica, California, started the design of its improved DC-1 (Douglas Commercial Model 1). No DC-1 production models were built, but the improved DC-2, with fourteen-passenger capacity, 750hp engines with controllable-pitch propellers, and wing flaps wiped out Boeing's initial lead

The Boeing Model 40 opened the era of combined passenger and mail service in 1927 by adding a passenger cabin to the traditional open-cockpit airmail biplane.

This four-passenger Model 40C belonged to Pacific Air Transport of California. Boeing and put the Seattle firm out of the twin-engine airliner business.

Reenter Lockheed

A newly reorganized Lockheed Aircraft Company (see chapter 2) reentered the airliner market in 1934 with the ten-passenger Model 10 Electra that became quite popular with several US and foreign airlines. It was more competitive with the Boeing 247 than with the Douglas DC-2 in spite of such later features as flaps and controllable propellers. To meet the Douglas and Lockheed competition, Boeing completed the last thirteen 247s as Model

247D with fully cowled engines, controllable propellers, and a cleaned-up windshield, but still refrained from using flaps. Most 247s already in service were sent back to the factory for conversion to Model 247D. But the improvements, which put the cruising speed of the 247D above the top speed of the basic 247, came too late to keep it competitive.

The Dominant DC-3

The DC-2 itself was surpassed in 1936 by one of the most successful airliners of all time, the Douglas DC-3. The initial version of the new model,

first flown on December 17, 1935, was the DST, for Douglas Sleeper Transport. This carried fourteen night passengers who were provided with Pullman-type sleeping berths. The major production model, the DC-3, carried twenty-one day passengers. Some 460 DST/DC-3s were delivered to the world's airlines by December 1941, making that design the most-sold airliner of that time. The only new twin-engine US airliners to enter service after the DC-3 were the fourteen-passenger Lockheed Model 14 and the fifteen- to eighteen-passenger Lockheed Model 18 Lodestar. Both were

The twelve-passenger Fokker F-10A was an improved American version of the ten-place European Fokker F-VII model, and was powered with three 420hp Pratt & Whitney Wasp engines. The Ford 5-AT trimotor was very similar except for having an all-metal structure. Museum of Flight

The final single-engine model on US trunk airlines was the eight-passenger 735–775hp Vultee V-1A. Cruising speed was 205mph, better than any of the contemporary twins. However, single-engine models were barred from trunk-airline service in 1934 for safety reasons. Gordon S. Williams collection

faster than the DC-3, but had less range and payload. The DC-3's unique combination of payload, cost, performance, and ease of maintenance enabled it to survive all twin-engine competitors and hold its position as a major piston-engine trunk airliner well into the 1960s.

The Four-Engine Era

Four-engine airliners appeared in Europe in the late 1920s. Western Air Express experimented with two four-engine thirty-two-passenger Fokker F-32s in 1930, but the market did not support them and they were retired within a year. The requirement for range and reliability dictated the use of four engines on long overwater routes, so Pan American World Airways (Pan Am) initiated the development of four-engine flying boats in 1930. Four-engine landplanes did not enter continental US airlines until 1940.

In 1935 Boeing introduced a then-radical high-speed four-engine bomber, the B-17, that did for the bomber business what the Model 247 had done for the airline business. It did not take Boeing long to develop a new pressurized thirty-three-passenger fuselage that used the B-17 wings, tail, and engines on the Model 307 Stratoliner. Pan Am and TWA ordered four and six, respectively. The first Pan Am model crashed on a test flight and the first TWA model was transferred to Howard Hughes, then TWA's major stockholder. TWA capitalized on the in-

The ten-passenger Boeing 247 of 1933 revolutionized the design of all subsequent airliners. This is the improved 247D model, introduced after the Douglas DC-2 and the Lockheed Model 10 Electra outperformed it and were more passenger-friendly. Boeing

The fourteen-passenger Douglas DC-2 with 750hp engines, controllable-pitch propellers, and flaps outperformed the Boeing 247 and won immediate worldwide airline orders. Design of the preceding DC-1 was initiated before the 247 flew. Airline service with the improved DC-2 began May 18, 1934. Gordon S. Williams

creased altitude capability of the pressurized 307 Stratoliner to inaugurate Over-the-Weather transcontinental service on July 8, 1940.

Even earlier, Douglas and several US airlines had foreseen the need for larger and longer-range four-engine landplane airliners. Five airlines and Douglas cooperated in the development of the prototype Douglas DC-4E, a design that could accommodate forty-two day passengers and be fitted with thirty berths for sleeper passengers. First flown on June 7, 1938, the DC-4E was route-tested by the airlines but was rejected as being too cumbersome and too costly to operate relative to its payload. Douglas then went back to the drawing board and designed a smaller, faster, and more efficient forty-passenger model that was also designated DC-4. The new model did not fly until February 14, 1942, by which time all airliner production had been taken over by the US Army. The DC-4 then acquired the Army designation of C-54 (R5D when used by the Navy). Airlines that had ordered the DC-4 had to wait until the end of the war for their factory-new DC-4s and converted war-surplus C-54s.

After TWA pulled out of the original DC-4E program, it joined with Pan Am again in getting Lockheed to revise and enlarge the fast four-engine thirty-passenger Model 44 Excalibur that Lockheed then had on the drawing boards. The result was the forty-four-day-passenger (thirty-passenger sleeper) Model 49 Constellation, originally called Excalibur A, which had the pressurization feature of the Boeing 307 and bigger engines that gave it a great speed advantage over the revised DC-4. However, as with the DC-4, the Army took over Constellation production as the C-69 and again the airlines had to wait until the end of the war for their new and war-surplus Connies.

Postwar Airline Operations

The war's end released hundreds of former military models onto the surplus market. Some airlines got back the DC-3s that the Army had taken from them in 1941-1942, and eagerly snapped up the military versions of the DC-3 that had been built for the Army as the C-47 and for the Navy as the R4D. Converted C-47/R4Ds were certif-

The twenty-one-passenger Douglas DC-3 (fourteen passengers in the DST sleeper version) entered airline service on June 25, 1936. Essentially an enlarged DC-2, it had the necessary combination of economics, performance, ease of maintenance, and a virtually age-proof structure that made it the world's most successful twin-piston-engine airliner. Gordon S. Williams

Lockheed was close behind Douglas in developing a fast four-engine airliner, the Model 049 Constellation. This is the prototype, photographed in both civil and military markings the day before its first flight on January 9, 1943. As with the DC-4, the airlines had to wait until the end of the war for their Constellations. Lockheed via Mitch Mayborn

icated and operated commercially as the DC-3C. The Army C-53 model was essentially a civil DC-3A and could be used commercially as such without recertification. The many military Lockheed Model 18 Lodestars that were released as surplus sold well, but did not return to significant airline use.

Four-engine transoceanic landplane operations began on October 23, 1945, when American Overseas Airlines (AOA) inaugurated service between New York and Bournemouth, England (near London), with a converted C-54 in direct competition with Pan Am's four-engine Boeing 314 flying boats. The trip required refueling stops at Gander, Newfoundland, and Shannon, Ireland. Nonstop New York–London service was still far in the

future. Pan Am inaugurated overseas service with Lockheed Constellations on February 3, 1946.

Both Douglas and Lockheed continued production of their DC-4 and Model 49 as civil airliners after cancellation of military contracts for their militarized equivalents. Since the Constellation was pressurized, more powerful, and faster than the DC-4, it gave Lockheed a two-and-a-half-year advantage over Douglas, the time that it took Douglas to develop and introduce its new higher-powered and pressurized DC-6, a stretched and upgraded version of the basic DC-4. Competition between Lockheed and Douglas would continue to the end of the piston-airliner era.

History of Lockheed Aircraft Corporation

The Lockheed Aircraft Corp. that introduced the ten-place Model 10 Electra airliner in 1934 was a brand-new company, established in June 1932. Lockheed's association with the design and manufacture of aircraft far preceded that date, however, going all the way back to 1912. This chapter does not attempt to catalog all Lockheed airplanes; however, it does show the early models that established the line, and key models that led to the development of the Constellation.

1912-1915

The present-day Lockheed Corp., parent company of what is now the Lockheed Aeronautical Systems Group, traces its origins back to the joint efforts of two brothers of Scottish ancestry, Malcolm (1887-1958) and Allan (1889-1969) Loughead (note spelling). The brothers combined their talents as pilot, mechanic, and airframe builder to produce a single-engine seaplane designed by Allan and called the Model G. To finance their efforts they formed the Alco Hydro-Aeroplane Co. The Model G was built and successfully flown in San Francisco, California, on June 15, 1913. The brothers bought out the other investors, but the firm was soon dissolved.

1916-1921

In 1916 the Loughead brothers formed a new company, the Loughead Aircraft Manufacturing Co., in Santa Barbara, California. There they built a twin-engine flying boat, the Model F. One employee, later to achieve fame on his own, was twenty-one-year-old John K. Northrop, a sharp high-school graduate who had been hired as a stress analyst.

The Model F was completed in March 1918, and was submitted to the US Navy for testing. The Navy did not buy it, but on the basis of Loughead Aircraft's demonstrated ability, gave the firm a contract to build single-engine Curtiss HS-2L flying boats which were needed at the time. However, only two of these were completed by the time the contract was canceled after the armistice.

The Model F-1 was then converted to a landplane, the F-1A, but was reconverted to a flying boat and used for passenger-hopping and whatever other work it could be hired out for. A small single-seat sportplane, the S-1, designed by Northrop, was an efficient little plane but did not sell in competition with less-expensive war-surplus models then flooding the market. With no customers, the Loughead Aircraft Manufacturing Co. went out of business in 1921. Malcolm capitalized on his automotive expertise to develop a four-wheel hydraulic brake system for cars and trucks and to found the Lockheed (note new spelling) Hydraulic Brake Company to manufacture it. Al-

lan entered the real estate business, and Northrop went to work for Douglas Aircraft in nearby Santa Monica.

1926-1929

In his spare time, Northrop designed a sleek high-wing cabin monoplane. While working on traditional boxy designs at Douglas, he had convinced himself that the key to improved performance was to get rid of as much of the built-in drag as possible. His dream ship was an all-wood cantilever monoplane—meaning no struts or wires—using the same molded-wood fuselage of oval cross section that he had developed for the S-1 to reduce drag.

Believing that Douglas, which was deeply committed to high-drag biplanes at the time, wouldn't be interested in the design, Northrop got in touch with Allan Loughead, who now spelled his name Lockheed. Allan was impressed with Northrop's design, and between them they founded a new organization, Lockheed Aircraft Co., to

The first Lockheed (then spelled Loughead) airplane was the Model G seaplane, launched in 1913. Powerplant was an 80hp Curtiss Model O eight-cylinder water-cooled engine. The Model G was finally scrapped in 1918. Lockheed

The second Lockheed airplane was the twin-engine Model F flying boat, powered with 160hp Hall-Scott engines. It was tested by the US Navy in 1918 but did not win a production order. Lockheed

In 1920 Lockheed introduced the single-seat Model S-1 sportplane, designed by John K. Northrop. With no suitable powerplant available, Lockheed had to design and build its own 25hp two-cylinder air-cooled engine. Lockheed

build it. With financial help from others, they set up shop in a small rented building in Hollywood, California. Their new airplane, designated the Model 1 Vega, started a long-lived practice of naming Lockheed airplanes after celestial bodies. The Vega, powered with a 220hp Wright J-5 Whirlwind engine, flew on July 4, 1927. It was sold to publisher George Hearst, who entered it in the infamous Dole Race from Oakland, California, to Hawaii.

The first Vega disappeared at sea, but its loss did not stop production. The firm moved into new quarters, a former brick works in Burbank, California. Vegas sold well to private owners and to small airlines. Northrop designed a low-wing variant, the Explorer, to use the same fuselage mold, which was subsequently used for other Lockheed model fuselages as well.

A minor variation of the Vega was the Model 3 Air Express. This was a prime example of a manufacturer trying to meet a customer's requirements. The customer, Western Air Express, bowing to the demands of pilots used to old-fashioned open cockpits located far aft on the fuselage, insisted on this location. In order to give the pilot visibility ahead, Northrop had to raise the wing above the fuselage to the parasol position, in effect making the Air Express a traditional biplane without a lower wing.

Western accepted the Air Express, but damaged it during testing and turned it back to Lockheed. With only seven built, the Air Express was not a commercial success. In the hands of such notable speed pilots as Frank Hawks and Roscoe Turner, however, it set many city-to-city speed records. Its major technical point of distinction is that it was the first commercial airplane to capitalize on the streamlining benefits of the new National Advisory Committee for Aeronautics (NACA) engine cowling that was to become so effective on subsequent Vegas and following Lockheed designs.

John Northrop left Lockheed in 1928 to start a new firm of his own, and Gerald Vultee took over as chief engineer. His first job was to finish the Explorer design, the second Lockheed airframe started but which had been set aside because of Vega priorities. Although an earlier design than the Air Express, the Explorer was identified as Model 4. Similar models were built as the Model 8 Sirius and the Model 8D Altair with retractable landing gear.

1929-1931

In July 1929, the giant Detroit Aircraft Corp. acquired 87 percent of Lockheed. The Lockheed factory remained in Burbank, but control was from Detroit. Engineers in Detroit introduced some innovations to established Lockheed designs, most notably a metal fuselage for nine Vegas that were designated DL-1, and one each Sirius and Altair. The fuselages were built in Detroit and then fitted with wooden wings, tails, and other components shipped from Burbank.

The Burbank firm continued to develop improved models, most notably the low-wing Model 9 Orion cabin monoplane. Introduced in April 1931,

the Orion was the world's fastest airliner at the time and enjoyed brisk sales even with the worldwide depression well under way.

In stifling the aircraft market, the depression hit Detroit Aircraft hard, and many of its aircraft manufacturing divisions that could not sell their products were forced to shut down. Lockheed was the only division still turning a profit, but it, too, was forced to shut down after the parent company

The first product of the new Lockheed Aircraft Co. was the Model 1 Vega. This is the first example, with designer John Northrop in the center of the group. Named Golden Eagle, the first Vega was entered in the 1927 Dole Race to Hawaii but was lost at sea. Lockheed

A special mail-express variant of the Vega was the Air Express. To give the pilot forward visibility from his aft-located cockpit, it was necessary to raise the wing from the top of the fuselage to the parasol position shown. Lockheed

15

went into receivership in October 1931. The Burbank plant carried on under banker's oversight until the following June, then was closed down.

1932 and On

On June 21, 1932, Robert Gross, an investment banker, and some other financiers bid $40,000 in US District Court in Los Angeles for the remaining assets of the Burbank organization, then appraised at $129,961. Their bid was accepted, and a new Lockheed Aircraft Corp., with many of the same employees and the same product line, started business in the same plant.

Vultee had left the old firm in 1930 to form his own company, and was replaced as chief engineer by Richard Van Hake, who assumed the same office in the new firm.

On-hand materials were used to complete four Vegas, one Altair, and seventeen Orions. Work on a new all-metal design, a single-engine transport similar to Vultee's new V-1, was started. Fortunately this was abandoned in favor of a ten-passenger twin-engine transport, the Model 10 Electra which continued the old model numbering system. In October 1934, the government ruled single-engine transports ineligible on the trunk airlines, so if built, Lockheed's single-engine design would have had no customers.

Electras with a variety of engines in the 400–600hp range found ready acceptance by the airlines and business owners and competed effectively with the Boeing 247. They beat the Boeing in sales as well, 149 to 75, but could not match the Douglas DC-2 with 193 total sales. The Electra entered airline service on August 4, 1934, and the last was delivered on July 18, 1941.

Further Airliner Developments

After the twenty-one-passenger Douglas DC-3 appeared in 1935, Lockheed introduced the Model 14 Super Electra in 1937. This was a fourteen-passenger twin-engine design that was essentially an enlarged and more powerful Model 10 and which offered 750–1,200hp engines. The 14's most notable new feature was the Fowler flap, a new type of wing flap that not only lowered to increase lift but extended aft as well to increase effective wing area. This detail was to carry over to the Constellation. Lockheed delivered 112 Model 14s between December 1937 and 1940, and an additional 119 were built under license in Japan.

Although the Model 14 was faster than the DC-3 it was smaller and not truly competitive. Prewar and postwar DC-3s dominated the twin-engine transport market after the war and the Model 14 was no longer part of the airline scene. The basic model enjoyed enormous success, however, after it was converted to a patrol-bomber for Britain's Royal Air Force under the name of Hudson. Most of the 2,941 built went to Britain starting in Febru-

Later versions of the Vega, like this Model 5B, featured 420hp engines under the NACA cowling first used on the Air Express, *and further improved streamlining by enclosing the wheels in "pants."* Lockheed

The Model 9 Orion combined the low wing and retractable landing gear of the Altair with the four–six place cabin and forward *pilot location of the Vega to create what at the time was the world's fastest airliner.* A. U. Schmidt

16

ary 1939, but some were diverted from British orders and served the US Army as A-28s and A-29s, and some were built for the Army as AT-18 bombing, navigation, and gunnery trainers. The US Navy acquired twenty Hudsons and designated them PBO-1.

In a move that surprised the industry, Lockheed, a firm that had no previous experience with high-performance military aircraft, introduced a radical twin-engine twin-boom fighter for the US Army, the XP-38. On February 11, 1939, this model set a transcontinental speed record which was followed by orders for a total of 9,924 by Lockheed and a further 113 built under license by Consolidated-Vultee. The P-38, which Luftwaffe pilots called *Der Schwanzlose Teufel*, or "Forked-Tail Devil," was to make significant contributions to the Constellation.

The Model 14 airliner was followed by another, the Model 18 Lodestar, which used Model 14 wings and tail on a stretched fuselage that could accommodate fifteen to eighteen passengers. But even with its greater speed, the Lodestar was still no match for the DC-3.

Lodestar sales to the airlines were moderate; most production went to the armed forces starting in 1941. Of 625 built, 419 were delivered to the US Army as C-56, C-57, C-59, and C-60, depending on the engine used. An additional thirty-five C-56s were drafted from civil owners. The US Navy also purchased ninety-five Lodestars and designated them R50-1–R50-6, and 121 were built under license in Japan. Military Lodestars came on the surplus market after the war, but none regained their prewar place as trunk airliners.

In 1940, Lockheed developed a patrol-bomber variant of the Lodestar for Britain, which named it Ventura. This was built in a larger and more modern factory that Lockheed built for its wholly owned subsidiary, Vega Aircraft Corp., on property adjacent to Burbank's Union Air Terminal.

With the Ventura Lockheed again developed a major military model out of a civil design, with some 3,028 built for Britain, the US Army as B-34 and B-37, and the US Navy as PV-1. The changes were far more extensive than those made on the Hudson. The basic

The first product of the reorganized Lockheed Aircraft Corporation was the ten-passenger Model 10 Electra that became a very successful airliner. Note only partial retraction of wheels. Extra space between two forward cabin windows accommodates the main wing spar, which passes through the cabin at that point. Gordon S. Williams

A larger fourteen-passenger transport was the Model 14 Super Electra introduced in 1937. The molded plexiglass nose cone covered the directional loop antenna. The most famous Model 14 was the one used by Howard Hughes to fly around the world in ninety-one hours and fourteen minutes July 10–14, 1938. Gordon S. Williams

17

Model 14/18 wings and tail were retained, but a new fuselage was used along with 1,850–2,000hp engines. These resulted in a new model number, 37, where the Hudson conversion of the Model 14 had been designated Model B-14L and -414. Gross weight of the Ventura I was 26,000lb compared to the Lodestar's 17,500lb. The final 1,600 PV-1s built for the US Navy (note the V for Vega in the designation) grossed 34,000lb, almost double that of the Lodestar.

Lockheed became a big-time manufacturer with the Hudson bomber version of the Model 14. A bomb bay was added under the cabin floor and a powered British gun turret was added on top of the fuselage. The airliner windows were retained. Imperial War Museum

The Model 18 Lodestar started the "stretched airliner" trend so popular today by fitting a longer fuselage to the existing Model 14 wing, tail surfaces, engines, and landing gear. Civil sales ended in 1941 and all subsequent production was for the armed forces. William T. Larkins

Lockheed Goes Four-Engine

With a market for larger transports in sight in 1938, Lockheed started the design of a four-engine airliner that went through several paper configurations before being finalized as the Model 049 Constellation. Airline contracts were signed and construction got under way, but World War II intervened. Thus the airlines that ordered Connies in 1940 did not get them until after the war.

The buildup for World War II, starting with the 1938 British order for Hudsons, raised Lockheed from a small firm that had only 332 employees in 1934 to one of the giants of the US aircraft industry. Employment in June 1941 was 18,724, and the wartime peak of 94,329 was reached in mid-1943. To meet its requirements for outdoor assembly area and aircraft parking and testing, Lockheed bought Burbank's Union Air Terminal in 1940 and renamed it Lockheed Air Terminal.

Principal wartime Lockheed products included the Ventura, P-38, and Boeing B-17 that was built in the Vega plant as part of the B.V.D. Pool of Boeing, Vega, and Douglas that had been formed to build the B-17 in quantity. Lockheed delivered 2,750 B-17Fs and Gs between mid-1942 and mid-1945.

The US Army also placed orders for military versions of the Model 49 Constellation as C-69 and took over those already under construction for airlines. The C-69 program suffered as an Army low priority, however; of 260 C-69s ordered, plus the nine taken from TWA, only eleven were delivered by war's end.

The airlines finally received their Constellations starting in late 1945. That product line went on to generate $1.5 billion in sales over the next thirteen years, after a late 1947 sales slump threatened to shut down the Connie production line.

Design of the Constellation

In 1938 Lockheed began studying designs for a new airliner with accommodations ranging from fourteen to forty-two passengers. The airplane that emerged from these designs was the Model 44 Excalibur, which could carry twenty-one passengers. Power was to be provided by four twin-row Wright Cyclone or Pratt & Whitney Wasp radial engines, giving a top speed of 241mph. Early discussions with Pan American (Pan Am) increased the accommodations to thirty and the top speed to 270mph and later increased them to thirty-six passengers and a speed of 300mph.

Mock-ups and wind-tunnel models had been built when TWA started looking for an airliner larger than those currently in use. TWA needed an airplane with nonstop, coast-to-coast capabilities that could fly above the weather. Specifications required that it be able to carry a payload of 6,000lb, have a maximum range of 3,500 miles without a payload, and a speed of more than 250mph at 20,000ft. Additionally, it had to be able to fit into existing maintenance hangars. Lockheed at first fought valiantly to convince TWA to purchase the Excalibur, but eventually reconsidered when Pan Am also became interested in this new proposal. The Excalibur project was dropped in favor of the new design, the Model 49 Excalibur A—ultimately renamed the Constellation—which was begun in the summer of 1939.

The specifications were largely set up by TWA's major stockholder, Howard Hughes, who was then at the peak of his piloting career, and Jack Frye, then president of TWA, who had done TWA's early over-weather flying in a specially equipped Northrop Gamma. Subsequent historians have given Hughes credit for designing the Connie. Not so.

He was instrumental, with Frye, in setting up the requirements for capacity, performance, and such details as cockpit layout. But credit goes to the leaders of the Lockheed design team, vice president and chief engineer Hall L. Hibbard, and the legendary C. L. "Kelly" Johnson, whose work for Lockheed started with the Model 10 Electra in 1933 and culminated with the triplesonic SR-71 Blackbird of 1964. Project engineer was Don Palmer.

Constellation Features

Layout of the Model 49 Constellation was conventional for the time: a low-wing monoplane of all-metal stressed-skin semi-monocoque construction. An innovation for a Lockheed airliner was the use of tricycle landing gear, a feature already seen on Lockheed's sensational XP-38 fighter. Another innovation, previously seen on the prototype Douglas DC-4E of 1937 and planned for the Excalibur, was a triple vertical tail. As on the Douglas, this was a product of necessity. A single vertical tail might have been more aerodynamically efficient, but with it the plane could not be housed in the standard hangars of the time.

Fuselage

The fuselage, a major contributor to the Connie's appearance, departed

The early version of the proposed Lockheed Model 44 Excalibur looked very much like a four-engine Model 10 Electra. It was 73ft, 2in long with a 95ft wingspan.

With 1,000hp engines the Model 44 was to have carried thirty passengers and a crew of four to six over 2,000 miles. Lockheed

from tradition in that the side-view centerline was not a straight line. It curved downward at the nose in a deliberate move to lower the structure there and shorten what would otherwise have been an overly long nose landing gear, while curving upward at the rear to raise the tail surfaces. The sweeping lines of the fuselage more accurately matched the actual flow of air along the body than those of other airliner fuselages, which were essentially long, straight cylinders with pointed ends.

Lockheed studied several different windshield and cockpit designs for the Constellation in an attempt to eliminate the windshield step. The first design featured a completely faired nose but was discarded when it was found to provide poor visibility. The second design was similar, but had the cockpit placed below the main floor and was soon discarded because of the unsuitability of the large nose. Next was a dual bug-eye design, with separate bubble canopies for the pilot and copilot, but it was found to be too claustrophobic and also caused increased drag. To counter this, Lockheed tried a single bug-eye but found that this also increased drag and created pressurization problems. A conventional step windshield was tried but the fuselage proved to be too wide, leaving the only solution to be a wraparound windshield with nine small panes.

The cross section of the fuselage was a circle, the most efficient shape for reducing the stresses of cabin pressurization. To further resist pressurization forces, the cabin windows, fourteen to a side, were circular instead of rectangular, and were quite small, only 16in in diameter. Maximum width of the 95ft, 2in fuselage was 11ft, 7½in at the leading edge of the wing, almost 2ft wider than the Douglas DC-4. Pressure bulkheads were located a foot aft of the nose and 12ft, 8in forward of the tip of the tail cone. Normal pressure differential was 4.1psi (pounds per square inch), giving an 8,000ft cabin altitude at an airplane altitude of 20,000ft.

Although normally considered part of the wing, the stub center section was built in as an integral part of the fuselage and was not removable.

Wing

The wing, with a span of 123ft (six more than the DC-4's), and an area of 1,650sq-ft (200sq-ft more than the DC-4's), was a direct scale-up of the P-38 wing. The airfoils were mixed: the reliable old NACA 23018 at the root tapering to the NACA 4412 at the tip. Aspect ratio was 9.17:1 and the planform taper ratio was 2.16:1.

The wings were built in two major assemblies on each side of the fuselage. The inboard panels supported the engine nacelles, fuel tanks, landing gear, and flaps, and were not normally removable. The removable outboard sections contained the fabric-covered ailerons and a removable wing tip.

Under Pan Am's supervision the Model 44 Excalibur grew to carry more passengers and began to look very much like an early Constellation. The design now had the *triple tail plus a deeper fuselage that was much like that of an enlarged Lodestar.* Lockheed

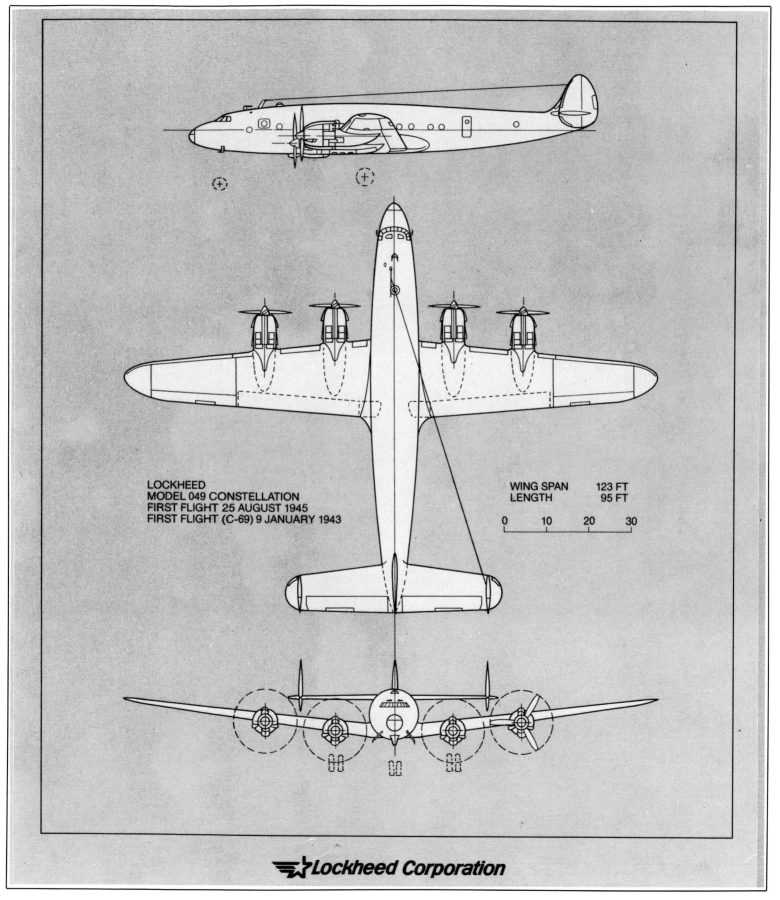

LOCKHEED
MODEL 049 CONSTELLATION
FIRST FLIGHT 25 AUGUST 1945
FIRST FLIGHT (C-69) 9 JANUARY 1943

WING SPAN   123 FT
LENGTH   95 FT

0   10   20   30

Lockheed Corporation

Three-view drawing of the Lockheed Model 049 Constellation, as built. Outlines and dimensions also apply to Models 649 and 749, the other short-fuselage Constellations. Lockheed

From the twenty-third airplane on, the ailerons were metal covered.

There were two main web-type spars in each wing, and the ribs between them formed the integral fuel tanks. Early Connies had two tanks in each inner wing panel, the inside one with a capacity of 780gal and the outer one of 1,555 for a total of 4,690gal. The smaller inboard tanks were H-shaped with a narrow area in the middle to make room for the retractable landing gear. This fuel could be dumped in flight to lighten the airplane for an emergency landing, but only from one side at a time. Later Connies had additional tanks in the outer wing panels and some in the 1049 series even had external tanks on the wing tips.

The Constellation was fitted with Fowler flaps, as pioneered by Lockheed with the Model 14. The wing flaps on either side of the fuselage were built in six separate sections, but when lowered appeared to consist of only two. There was a short section under the stub wing and five separate but linked sections on the inboard wing panel. During the first half of flap extension, the motion was aft to immediately increase the effective wing area, but the last half of the extension took the flaps

The circular cross-section fuselage of the Constellation was not built up as a single assembly; it was assembled from eight separate short sections such as these. Lockheed via San Diego Aerospace Museum

to the full-down position of forty-two degrees. All Connies could be fitted with pneumatically operated Goodrich rubber deicer boots on the leading edges of the wing and tail surfaces.

Tail Surfaces

The three-fin-and-rudder combination carried on the Lockheed structural trademark that was first introduced on the Model 10 and finalized on the Model 14—an eccentric ellipse with part of the area located below the horizontal tail. The three-fin arrangement on the Connie, instead of Lockheed's customary two, owed much to the

Douglas DC-4E and met TWA's requirement that the airplane fit into existing hangars. The horizontal tail inherited a detail of all twin-engine Lockheeds back to the Model 10, the P-38 included. This was a small portion projecting outboard of the outer fin and rudder on each side. While this added effective area, the question remains whether this benefit was outweighed by the additional intersection drag. The three rudders and the two-piece elevator were fabric covered.

The outboard rudders were interchangeable, but the center one was not. The two elevators were interchange-

able right and left, which resulted in an oddity: the tab hinge on the right elevator was on the bottom, while it was on the top of the left elevator.

Landing Gear

Tricycle landing gear had been used on only one other airliner before the Connie—the Douglas DC-4E of 1937 and the redesigned production DC-4 that was not to fly until February 2, 1942. The Connie gear was similar, except for having two nose wheels instead of one. The main gear also had two wheels, each with hydraulic brakes, and were retracted forward hy-

The stub wing and center section of Constellations through Model 1049 was built as an integral part of the fuselage. Here an inboard wing panel is about to be at- *tached to the fuselage of a Model 1049/ R70-1 for the US Navy. Lockheed via San Diego Aerospace Museum*

Head-on flight view of C-69 43-10317 shows flat center wing stub and sharp *dihedral angle starting at inner ends of the inner wing panels.* Lockheed

Front view of a Constellation with Fowler flaps fully lowered. Outer flaps are in five separate sections; the inner section is on *center wing stub and partly under the fuselage.* Bowers collection

draulically while the nose gear retracted aft.

If the hydraulic system malfunctioned, the nose gear couldn't be extended by the normal slipstream free-fall gravity method. To alleviate this problem, Lockheed installed an electrically driven standby pump for emergency extension. All hydraulic controls were equipped with manual override, and the airplane could easily be handled in the event of a complete hydraulic system failure.

Powerplant

The engine initially selected for the Constellation was the new and relatively untried Wright R-3350 that had so far been flown only in a few prototype airplanes—the Martin XP2M-1 and Consolidated Model 31 flying boats, and the Douglas XB-19 that first flew in June 1941.

The full designation of the engine in the first civil Connie was R-3350-745C18BA-1, but for the civil models most references delete the R-3350 part. It was used by the military, however, without the following numbers and letters. A sequential dash number was used instead, such as R-3350-34, -35, and so on, with odd dash numbers identifying US Army engines and even dash numbers identifying US Navy engines.

In its early production form, the twin-row eighteen-cylinder R-3350 delivered 2,000 continuous horsepower at 2,400rpm and 2,200hp for takeoff (two minutes) at 2,800rpm on 100/130 octane fuel. The propellers were three-blade Hamilton-Standard hydromatics

Takeoff view of a Model 749-79-22 Constellation in TWA markings shows the nose landing gear retracting aft and the main landing gear retracting forward. Bowers collection

Close-up of the unique triple tail of Constellation Models 049 through 749. This is a Model 749A acquired by Pacific Northern Airlines from original owner TWA. Its five-segment outer wing flap and single-segment inner flap are visible here. Peter M. Bowers

with full feathering. Engine accessories powered the airplane hydraulic system, the cabin pressurization system, and the 28vdc electrical system.

Originally the engines in the Constellations were fitted with conventional carburetors, but chronic trouble with engine fires in Connies and Boeing B-29s that used the same engine resulted in a change to fuel-injected engines following the grounding of all civil Connies on July 12, 1946, for other problems.

Because of the large propeller diameter, 16ft, 10in, it was necessary to gear the propellers well below engine speed for efficiency. The expression of the ratio itself is a bit confusing, being expressed either as whole numbers, such as 16:7, or as fractions, 0.355:1, sometimes in the same document!

A big jump in takeoff power from 2,800hp to over 3,400hp followed the addition of turbo-compound devices to the engines of later-model Connies. Three turbines, each driven by the exhaust of six cylinders in the manner of a turbosupercharger, were connected through shafts to the engine crankshaft to deliver over 500 additional horsepower per engine.

In another attempt to reduce drag as much as possible, the Constellation was originally designed with completely enclosed engine nacelles. The air intake was located at the nacelle-wing junction and then turned 180 degrees to the front of the nacelle where it turned another 180 degrees to pass over the engine. Wind-tunnel tests in-

Constellation power packages were tested in a modified Lockheed Ventura bomber that was nicknamed Ventillation. *Its shortened nose cleared the larger-diameter pro-pellers. Wartime camouflage covers air terminal and hangars in the background.* Lockheed

dicated that this system contributed only negligible savings over normal nacelles because of the necessary bends. Ultimately, long-chord tapered cowlings were used on the Constellation and the entire powerplant package was completely detachable.

Engine nacelle configurations were tested on a Ventura that had been kept at Lockheed as a test plane. This airplane, the seventh Ventura I on the original British contract, was called *Ventillation* in a combination of the two airplanes' names.

Controls

Something new was added to airliner control systems with the Connie hydraulic boost. Because of the airplane's size and speed, hydraulic boost would greatly reduce the pilots' workload. However, this pioneering created a problem: the designers didn't know what stick forces should be used. To solve this they queried numerous pilots to determine what was acceptable. The final solution was a boost ratio of 9.33:1 (the pilots' effort boosted 9.33 times) for the elevators, 23:1 for the rudders, and 26:1 for the ailerons.

The hydraulic actuators were connected to each control surface by push-pull tubes and the pilots' controls were connected to the actuators by cables. The ailerons had a high degree of differential, with twenty-five degrees of upward movement and only ten degrees downward movement. The elevators had forty degrees of upward travel and twenty degrees down, while the rudders moved thirty degrees in both directions.

Accommodation

As originally laid out for the airlines, the Connies provided first-class seating for sixty passengers in twelve rows of seats, two seats on the right side of an 18in aisle and three seats on the left, with a pitch, or distance between seats, of 38in. Later, when high-density coach seating came about, the short-fuselage Connies could carry up to eighty-one passengers. High-density seating for the later stretched models was ninety-nine. Early airline models had two dressing rooms with toilets at the rear of the cabin, while follow-on models had one or two additional toilets forward. Early models did not have a galley as such. Later ones had what were originally called pantries and, later, full-service galleys, near the rear of the cabin on the right-hand side.

One baggage compartment and a separate cargo compartment, with both external access and access through the cabin floor, were located ahead of the wing, beneath the floor; another pair of compartments was located behind the wing.

As designed, the Constellations were to have optional sleeping berths, up to twenty-two berths plus four seats in the maximum sleeper version. Be-

The interior of the Model 049 Constellation was Spartan by today's standards but was up to the airline standards of the time. The two most popular interiors were the forty-seven-passenger overocean and the fifty-one-passenger overland. Only American Overseas Airlines (AOA) ordered the forty-three-passenger interior. Pan Am

cause of the military takeover, this feature was deferred until the Model 749 of 1947. Even on Connies so equipped, the full possible number of berths was seldom used.

Many believe that when Lockheed was designing the cockpit, it grouped the necessary instruments, panels, seats, and controls logically on a well-proportioned part of the mock-up. After everything was laid out, it was shoved into the smallest space possible and called the cockpit. At best, the Constellation's cockpit could be defined in one word—tight. After one look into the cockpit from the cabin side, the first question was, How do you see out? It actually wasn't bad, since the pilots were seated only a few inches from the windshield and the nine-pane design did provide exceptional vision.

The flight deck was laid out for two pilots with dual controls. A flight engineer's station behind the copilot faced outward. His or her position included engine controls for the carburetor air, throttles, engine superchargers, mixtures, and fuel shutoffs. With most of the engine controls operated by the flight engineer, the pilots' workload was greatly simplified during takeoff and landing. Opposite the flight engineer and facing the rear was the optional radio operator. Behind a bulkhead that separated the cockpit from the rest of the cabin was the optional navigator and a crew rest area, since on very long flights it was necessary to carry a complete relief crew.

The crew boarded the airplane through a door in the right side of the nose, while the passengers boarded

The close confines of the Model 049 Constellation cockpit can be seen clearly in this photo. The center pedestal contains the throttles, flap controls, and the elevator trim wheel. Both pilots have complete flight instrumentation. Pan Am

through a single door on the left side at the rear of the cabin. Later cargo and convertible passenger-cargo models had two cargo doors on the left side, one ahead of and one aft of the wing. These large upward-hinged doors contained smaller doors for personnel entrance.

Constellation Designations

The Connie originated as Lockheed Model 49 (L-49 in most references) and appears as such in Lockheed and Federal Aviation Administration (FAA) documents. However, the two-digit model number appears in most references as 049, and subsequent developments advance in increments of 100: Model 149, 249, and so forth. Thus, we will use 049 in further references for the sake of model consistency.

The basic model number was followed by a dash and a two-digit number that identified the particular series of the R-3350 engine installed. For example, -46 identified a 745C18BA-3 version of the engine while -79 identified the 749C18BD-1 version. A second set of two-digit numbers identified various cabin and equipment details. FAA documents disregard the last two digits, but they appear in many Constellation listings. Significant changes within a basic model were identified by letter suffixes as Model 1049A, 1049B, and so on.

Constellation Serial Numbers

As with all airplanes, the Connies had factory serial numbers, called c/ns for constructor's numbers by airplane data buffs. These started with 1961 for the first Connie and ran consecutively

to 1980 for the twentieth. From there continuity was broken, with Model 049s starting again at 2021 and continuing to 2088, the last 049 built. Occasional breaks in continuity were the result of airframe cancellations, but often the introduction of a new model started a new c/n run, as in the Model 749 starting at 2501 and running to 2590 with the block of numbers 2518–2524 used for seven Model 649s. Model 749As started at 2600.

Constellation Models

Starting with the Model 049 (or 49), the designated developments of the Constellation progressed in increments of 100 to Model 1649. Not all were built, but since some of the unbuilt models influenced subsequent models, the full list is presented here:

Model 049 Initial model as designed for and ordered by airlines. Early articles completed and delivered to the US Army as C-69s. Most C-69s were reconverted to 049s for airlines; late 049s were built as such. See chapters 4 and 5 for details.

Model 149 None built as such. A 1940 proposal called for a lightened Model 049 with 875hp Wright Cyclone R-1820-63 engines. Applied in 1947 to Model 049s fitted with Model 749 wing panels containing additional fuel tanks.

Model 249 Study for conversion of Model 049 to a high-speed heavy bomber in competition with Boeing B-29, Douglas XB-32 (not built), and Consolidated B-32. Designated XB-30 by the US Army, but did not get off the drawing board.

Model 349 Original designation for thirty long-range versions of Model 049 ordered by Pan Am in 1940. Taken

over by the US Army as C-69B, but no airplanes built as such.

Models 449 and 549 None built. Proposed improvements of Model 049/C-69 for post-World War II airline use. Designations replaced by Model 649.

Model 649 Improved Model 049/C-69 developed for airline use. See chapter 6.

Model 749 Long-range Constellation with additional fuel in outer wing panels. See chapter 6.

Model 849 None built. Was to have been the basic Model 749 fitted with 3,250hp turbo-compound Wright R-3350 engines as used in the later Model 1049C.

Model 949 Proposed high-density passenger model that could be converted to all-cargo by installing hinge-up cargo doors ahead of and behind the wing. Not built, but details adopted on some later models.

Model 1049 Stretched-fuselage Super Constellation. See chapter 8.

Model 1149 None built. Study to replace reciprocating engines of Model 1049G with Allison turboprop engines.

Model 1249 Design goal of Model 1149 achieved with installation of turboprop engines for four US Navy R7V-1s that were redesignated R7V-2 and YC-121F. See chapter 9.

Model 1349 Designation deliberately skipped.

Model 1449 None built. Study to lengthen Model 1049G fuselage 55in, install new wing, and use turboprop engines.

Model 1549 None built. Further study to lengthen Model 1449 by 7ft, 11in.

Model 1649 Final Constellation model with entirely new wing on the Model 1049G fuselage. See chapter 10.

World War II and the C-69

World War II, which started in Europe in September 1939, had a major effect on the US aircraft industry even before it started. French and British orders for existing and soon-to-be-developed US airplanes did much to expand US aircraft manufacturing facilities. Lockheed was a major beneficiary of this buildup as a result of its quick conversion of the Model 14 airliner into the famous Hudson bomber. The Hudson orders financed a major expansion of Lockheed's Burbank factory long before the United States was drawn into the war by the Pearl Harbor attack of December 7, 1941.

Even with large military orders in hand, commercial business carried on at Lockheed, with the latest airliner under development being the Model 049 Constellation.

Initial Orders and Procurement

TWA placed the first firm order for nine Model 049s early in 1940. This was later increased to forty, with Pan Am coming aboard with an order for an additional forty, ten of them 049s and thirty longer-range Model 349s. The Dutch airline Koninklijke Luchtvaart Maatschappij (KLM) ordered four more for a start-up total of eighty-four.

However, this was not to be a simple case of taking a customer's order and deposit, building tooling, and ordering material. US involvement in World War II was imminent, and because of the big buildup of US aviation and manufacturing for both the United States and the Allies, the government had set up a Commercial Aircraft Priority Committee that had the final say as to who could build how many of what airplane model.

By early 1941 the Army was beginning to draft civil airliners right off production lines, and it was soon apparent that the commercial debut of the Constellation would be postponed indefinitely by the US Army takeover. In fact, the Connie looked so good to the Army Air Corps that it ordered 180 for itself under the designation C-69B, while assigning the designations C-69 and C-69A to the civil models already on order. As of May 4, 1941, Constellation orders were as follows:

Model/Destination	Quantity
049-46-10 for TWA; to become C-69	40
049-46-10 for Pan Am; to become C-69	10
349-43-10 for Pan Am; to become C-69A	30
349-43-11 for the US Army; to become C-69B	180

KLM had dropped out of the program by this time and the distribution was approved by the committee. The Army would own the airplanes, with the airlines operating them under contract to the Army.

The Army C-69 Program

Use of the series letter *B* for the Army's first contracted Connie indicated that earlier Army versions of the aircraft had already been identified. Since the Army was then drafting new civil airliners right off the factory floor, it looked like the service planned to do the same with some of the yet-to-be-built Connies. Such was the case. The fifty 049s destined for TWA and Pan Am were designated C-69 and Pan Am's thirty 349s were designated C-69A.

The first flight of the Lockheed Constellation, which the US Army had taken over and designated C-69, was from Lockheed Air Terminal in Burbank on January 9, 1943. Lockheed

The first C-69 in flight. Civilian registration NX25600 was painted on the day before the first flight, but the Army tail number 310309 was excluded. Use of company trademarks seen on nose and tail was very unusual on a military airplane. Lockheed

The first C-69 was soon stripped of its Army camouflage for continued Lockheed flight testing and delivery to the Army.

This example does not have the navigator's astrodome that was fitted to all other C-69s. Lockheed

These models were both passenger transports, not combat cargo types as required for military airborne task forces. However, during later negotiations with the Army, the 349s were to become cargo-plane designs with reinforced floors and a large cargo door on the left side of the fuselage, at the rear of the cabin. The C-69Bs were also to be cargo planes, but with gross weight increased from the C-69A's planned 67,000lb to 86,000lb.

This was not to be an easy transition, as Lockheed had no design or construction experience with such a large type. In December 1941, the Army initiated action for Lockheed to obtain military cargo-plane design and equipment data from Douglas. That firm had built the Army's first dedicated cargo plane, the C-1, in 1925, and had produced highly successful cargo versions of the civil DC-2 airliner as the C-33 and the C-39. It was then doing the same thing with the DC-3 to produce the C-47 and was well into the even larger DC-4/C-54. Lockheed obtained the required data and applied it to the redesign of the civil Connie as a military cargo plane.

Initial C-69 Configurations

Over its troubled production life the C-69 underwent many changes of Army thinking as to its mission and the equipment appropriate to that mission. As proposed in the 1941 negotiations and firmed up in July 1942, the configurations of the Army's C-69s at that time were as follows:

C-69 Basic airline airframe with the interior stripped of civil furnishings and the original gross weight of 65,000lb. It was reequipped as a military transport with forty-four seats on the right-hand side, four folding four-place benches on the left, and two lavatories in the rear. Military radio included SCR (Signal Corps Radio) -274 command radio, SCR-287 liaison set, SCR-269 radio compass, RC-43 marker beacon, RC-36 interphone, SCR-211 frequency meter, SCR-535 IFF (identify friend or foe) equipment, SCR-578 emergency transmitter, provisions for Lorenz 1124A and 1125A equipment, and RC-103 localizer receiver.

C-69A Similar to the C-69 except for the added high-density seating for 100 troops and a gross weight increase to 67,000lb. The cabin windows were to be fitted with rifle grommets as used on the Douglas C-47 and C-54 for defensive fire by the troops. A large upward-hinging cargo door was to replace the standard passenger door. No C-69As were built.

C-69B A long-range transport with the tankage of the Pan Am Model 349-43-11 and gross weight increased to 86,000lb. Otherwise it was similar to the C-69A with provision for 100 troops. No C-69Bs were built.

Later C-69 Configurations

The Army's vacillation with the C-69 program resulted in the cancellation of the C-69A and C-69B designa-

The second C-69 shared with the first the exclusive detail of small upper-berth portholes above the main cabin windows. This was also the first C-69 with a navigator's astrodome. A single landing light was located inside the lower transparent portion of the nose. Lockheed

tions and their replacement with two new ones while the first C-69 was diverted to a test program and redesignated.

C-69C The eleventh C-69, actually the ninth C-69-1, Army serial number 42-94550, was equipped as a VIP (very important person) transport with forty-two deluxe passenger seats, additional cabin windows, and 2,200hp R-3350-35A engines. An order for forty-nine more C-69Cs was considered by the Army but rejected. It was leased by the Army to TWA as a pilot trainer. TWA operated it under civil registration NX54212 until July 30, 1946. The Army then declared it obsolete, redesignated it ZC-69, and sold it to British Overseas Airways Corporation (BOAC) in April 1948. BOAC later had it upgraded to Model 049E standard in May 1953.

C-69D Another proposed passenger version with fifty-seven seats but a reported gross weight of 100,000lb. A major change was to replace the rubber deicer boots on the wings and tail with a thermal anti-icing system. Three C-69Ds were ordered, but none were built.

XC-69E The first C-69, serialed 43-10309, was used as a test bed for 2,100hp Pratt & Whitney R-2800-83 engines. This one airplane was fitted with the thermal anti-icing system proposed for the C-69D. Sold late in 1945 to Howard Hughes, who flew it under its original civil registration NX25600, Hughes sold it back to Lockheed in 1949, and Lockheed modified it extensively as the prototype of the Model 1049 Super Constellation program and the Navy WV-2 program with the new civil registration NC9700.

C-69 Factory Designations

In 1942, with several manufacturers building the same airplane—Boeing, Vega, and Douglas were all building the B-17—and manufacturers expanding into several widely separated plants, the Army assigned two-letter abbreviations to the end of the airplane's designation to identify the plant in which it was built. The C-69, for instance, was identified as the C-69-LO for the main Lockheed plant at Burbank. Since all C-69s were built there, use of the -LO suffix is redundant for reference to the C-69 and later Army and Air Force Constellations and therefore will not be used in this book.

C-69 Block Numbers

Early in 1942, when many changes were being made to Army airplanes while they were still on the production line, the Army instituted a block number system whereby small changes that did not justify a designation change, as from C-69 to C-69A, could be identified and tied to specific airplanes. The system started with -1, as C-69-1, then proceeded in blocks of five—C-69-5, C-69-10, and so on. The intervening dash numbers were used by modification centers that made changes of their own. A C-69-5 with a certain postdelivery change could become a C-69-6, for example.

The first two C-69s were just that, with no dash numbers. The next seven were C-69-1 while the remainder became C-69-5. Some records indicate

While the second C-69 was on loan to TWA, Howard Hughes had it painted in full TWA markings with only the Army tail number 310310 *between the TWA tail stripes.* Lockheed

The second C-69 after delivery to the Army. Note Air Transport Command (ATC) *logo on rear fuselage and oversize star-and-bar insignia.* Lockheed

that the last two of the twenty-two C-69s delivered to the Army were sixty-three-passenger C-69-10s, but Army documents do not verify this block number.

US Army Serial Numbers

Aircraft identification for the photos in this book and others will be simplified by an understanding of the US Army and Air Force serial numbering system. Since 1922, Army airplanes have been procured under federal fiscal year budgets. The first airplane contracted for in fiscal year 1941 (July 1, 1940 through June 30, 1941), for example, was serialed 41-1. When applied as a "tail number," use of which started in January 1942, the first digit of the fiscal year was deleted, along with the dash. Since it was used as a radio call sign, the tail number was expanded to four digits, appearing as 1001. Later, long after wartime numbers had gotten into the six-digit range (seven, when using the fiscal year digit), the tail number was cut down to five digits starting at the left even when that meant deleting the fiscal year digit.

Production Starts

In March 1942, it was decided that the Army would buy all of the Connies then on order to give it better control of production and inspection. However, the Army had a procurement dilemma: should it buy the airplanes from the airlines, which already had deposits down and legal identity with the production articles, or buy them from Lockheed, which could still legally sell them? With the airlines still involved, title was becoming confused, with TWA, Pan Am, and some US Army Ferry Command (AFC) inspectors all clamor-

Because of all the fixes required to make the C-69s acceptable to the Army, Lock- *heed set up an outdoor modification line at the Lockheed Air Terminal in Burbank.* *C-69-1 43-10315 is in the foreground. Lockheed*

ing for ownership. The Army then cracked down, informing Lockheed that the Air Materiel Command (AMC) at Wright Field was the procuring and controlling agency, and to disregard the airline people and even the Ferry Command inspectors.

On June 14, 1942, the existing contracts were revised whereby the Army could buy the original nine 049s from TWA, then under construction by an Army 1943 fiscal year contract, and lump the others with the 180 already ordered as C-69B by a fiscal year 1942 contract that would then become effective. Orders for even more C-69s came to a total of 313.

Such totals were never attained, however. Construction on only twenty-two C-69s was started as such, and the Army accepted only fifteen of those. Even with a clear need for combat cargo planes in 1942, the C-69 program never really got off the ground.

Lockheed was told to concentrate its efforts mainly on P-38 production and its B-17 program. A major benefit to Lockheed from the B-17 program, however, was experience in the mass production of large four-engine aircraft.

After much renegotiation and exchange of correspondence, the C-69 procurement program was revised on July 14, 1942, some six months before the prototype flew, as follows:

Army Model No.	Lockheed Model No.	Quantity	Gross Weight (lb)
C-69	49-46-10	20	65,000
C-69A	49-46-11	30	67,000
C-69B	349-43-11	210	86,000

Contract Changes

A new contract with Lockheed in September 1942 reduced the number of standard C-69s to eleven; the nine 049-46-10s then being built for TWA

were to be purchased from the airline rather than from Lockheed. Since the Army contract with TWA was signed in the government's 1943 fiscal year, these nine C-69s got fiscal year Army serial numbers 43-10309 through 43-10317. The other Army Connies, having been contracted for earlier with Lockheed, were given fiscal year 1942 serial numbers, starting at 42-94549. The cost of the revised 251-plane Lockheed contract was now $155,357,608.44, plus a fixed fee of $7,767,880.42. (In view of today's megabuck contracts ending in many zeroes, one wonders how such odd figures, down to fractions of dollars, were arrived at.) At any rate, each aircraft cost approximately $650,000.

First Flight and Early Testing

The first Constellation was flown on January 9, 1943. At the time, the plane carried full US Army camouflage

C-69-1 43-10314 in flight. The cabin window pattern differs from the first two C-69s, and the upper-berth portholes have been deleted. ATC logo appears on rear fuselage. Lockheed

and US military star insignia but no Army serial number on the tail. During taxi testing, it also carried civil registration NX25600 in white over the camouflage, and the Lockheed trademark was on the tail. These numbers were painted over before the first flight. Also, the Lockheed trademark was painted on the nose, and an oversize US Army data block was painted on the left side of the nose below the trademark. Oddly, the standard Army tail numbers were not applied. Use of company trademarks on a military airplane was very unusual.

The pilot was Edmund T. "Eddie" Allen, premier engineering test pilot in the country, who was on loan to Lockheed from Boeing for the occasion. (Boeing had also loaned him to Curtiss-Wright for the test flight of the CW-20 airliner that became the prototype of the Army's C-46 Commando transport.) Allen would later die in the crash of a B-29 due to an unextinguishable fire started by an R-3350 engine. Copilot was Lockheed's Milo Burcham of P-38 fame.

Six flights were made the first day, but accounts differ. Some say that the first five were made at Burbank, with the sixth landing at Muroc Army Air Base, California, while others say that the first landing and all subsequent flights were made at Muroc. Whatever, the prototype was soon returned to Burbank, stripped of its camouflage, and used by Lockheed to complete Phase One (manufacturer's) flight test-

The fourteenth C-69, a dash five, serialed 42-94553, was loaned to Lockheed for use as a demonstrator to the airlines. Here it carries Eastern Air Lines markings and the fictitious registration number NX70000. Eastern eventually bought fourteen Model 649 Constellations. Lockheed via San Diego Aerospace Museum

ing, and by TWA and Pan Am for further testing and publicity before being turned over to the Army in July.

The customer airlines, most notably TWA, were involved in testing the original C-69 and several others, after the Army owned them. Howard Hughes and Jack Frye, with some distinguished passengers, made a record nonstop flight from Burbank to Washington, DC, in the second C-69, Army serial number 43-10310, on April 17, 1944. Their elapsed time was six hours, fifty-eight minutes. The story that the flight had to be made in record time in order to reach Washington before too much gas was lost from the badly leaking tanks doesn't hold up in view of statements that most of the flight was made at only 65 percent power.

With this, Hughes pulled off a real publicity coup and demonstrated his clout in the industry. While the airplane was Army property on loan to him (as TWA), he had TWA markings painted on it, but no civil registration or military insignia, only the Army tail number 310310, between the trademark TWA red stripes on the fins and rudders.

Engine Problems

Even before the first C-69 flew Lockheed was facing up to the serious problems with the Wright R-3350 engine that had already appeared in other aircraft programs, notably the B-29. In December 1942, Lockheed requested permission from the Army to begin studies for substituting the 2,100hp Pratt & Whitney R-2800 engine for the R-3350 of the C-69. Because of the generally similar size and configuration of the two, installations in the airplane would be interchangeable. The Army authorized the studies, but did not assign a revised airplane designation at the time.

On February 20, 1943, less than two weeks after the C-69 flew, the AMC at Wright Field ordered the C-69 grounded because of R-3350 engine problems. Several such orders were issued during the year, with the result that the first C-69 was airborne for only three months of 1943.

Lockheed was bitter toward the Wright Engine Division of Curtiss-Wright over the R-3350's chronic problems. Most seemed to stem from the ignition system, which Lockheed accused Wright of never having tested properly in laboratories or in flight. Another serious problem was the unscrewing of the cylinder head from the cylinder base in flight, which caused violent backfiring and in-flight fires.

Wright came up with a new type of ignition harness, which overcame many of the ignition problems, and made the fixes that eventually turned the R-3350 into a reliable engine. However, the early break-in troubles were a major headache for both the Boeing B-29 and the Lockheed C-69 programs. Lockheed rightfully claimed that the engine problems were a major factor, along with being a low Army priority, in slow progress of the C-69 program.

Other roadblocks for the C-69 program included higher priority given to the B-29 program in the allocation of R-3350 engines, plus the frequent grounding of the test C-69s due to R-3350 engine problems, with most of the groundings ordered by the Army as the result of troubles encountered with the B-29.

Production Cutbacks

Between the C-69's engine problems and its low-priority status, Lock-heed protested that it could not meet the original contract delivery schedules. The first C-69 was supposed to fly on August 31, 1942, but did not get airborne until the following January 9. The second airplane was to have been ready for the Army's Model 689 board inspection on March 17, 1943, but the inspection didn't begin until May.

Because of these delays, plus engine shortages and the need to renegotiate contracts due to cost changes resulting from various changes and fixes, Lockheed informed the Army that the first C-69B could not be delivered until early 1945. The Army accepted this statement reluctantly, and took a serious look at the whole C-69 program.

The C-69's advantages over the Douglas DC-4, which the Army had taken over as the C-54, were largely negated by the fact that Douglas had made a quick change on the C-54, built as a civil airliner, to the C-54A, a dedicated cargo plane with cargo door, reinforced floor, and full military equipment. By May of 1943, when the Army reevaluated the C-69, the C-54A was already starting to make airlift history.

It was then that the Army recommended the C-69 program be cut back. Only the first twenty planes on the TWA and Lockheed contracts would be com-

The one-only C-69C-1, serialed 42-94550 and the eleventh C-69 built, was leased to TWA for use as a crew trainer in mid-1946. It was later converted to a Model 049 airliner for British Overseas Airways Corp. (BOAC), which had bought it from the Army. A. U. Schmidt

pleted. However, the cut was not that drastic at the time. Production of the C-69A would be canceled; production of basic C-69s was to continue beyond the initial twenty, and engineering was to continue on the C-69B, but no tooling was to be started.

On June 3, 1943, the Army and Lockheed agreed that C-69s would be built in quantities not to exceed ten per month through 1944 and that only three C-69Bs would be built, and those with minimum tooling. By June 12 Lockheed had outlined a production schedule that totaled ninety-eight

C-69s before the end of 1944 and requested tooling and material allocations on that basis. The Army's Production Engineering Section ruled that this schedule was unsatisfactory and cut it back to seventy-nine airplanes. By June 24, 1943, the existing contracts were revised to eliminate the 210 C-69Bs; 207 airplanes were to be delivered as standard C-69s, and there were to be only three C-69Bs.

Production and Service Problems

As a completely new design, the C-69 had all of the initial bugs common

to most new planes, plus the separate problems of a new and unproven powerplant. On July 12, 1944, the right main landing gear of the third C-69 collapsed during a taxi run. This resulted in design of a new forging and further delay to the program.

Because of accelerated wartime production demands, some of the necessary fixes were made after the planes left the production line. When engineering and tooling allowed, fixes on later airplanes were made at the plant. In June 1945, the Army directed that C-69s needing changes off the produc-

After four previous C-69s were rejected by the Army and one used for static test, the Army accepted C-69-5 42-94558 as its fifteenth and last C-69, then leased it back *to Lockheed for tests of the Speedpak shown. Compare window pattern to the C-69C-1. Lockheed*

tion line were to be sent to a modification center in Dallas, Texas. However, the C-69s could not be accommodated there because of the increased workload, so Lockheed set up an outdoor modification line at the Lockheed Customer Service Center at Lockheed Air Terminal in Burbank.

The C-69's problems were so numerous and serious that in June 1944, the Army directed that the few then in service test status as a joint effort of Air Service Command (ASC), Air Ferry Command (AFC), and the new Air Transport Command (ATC) were to be tested by military personnel only, and that the airplanes were not to be flown outside of the United States until the authorities were convinced of their safety. Altogether, 486 different modifications were found necessary for the C-69.

The C-69s were eventually cleared for flights out of the country. A C-69-5, Army serial number 42-94551, made the first ever transatlantic crossing by a Constellation on August 4, 1945, when it flew nonstop from New York to Paris in a record time of fourteen hours, twelve minutes. Its career was short-lived, however, since it was the first Constellation involved in a major accident. The C-69 was being operated by TWA for the USAAF when the accident occurred on September 18, 1945. It was eastbound over Kansas when an engine fire developed that couldn't be extinguished. The crew made a belly-landing in a cornfield near Topeka without injury, but the airplane was engulfed in flames and became a total loss. All C-69s were grounded for further repair. As a result, Lockheed proposed fifteen additional safety modifications, nine of them mandatory, before returning to flight status.

Further Cutbacks and Contract Cancellations

Because of the low priority of the C-69 program and its R-3360 engine problems and shortages, only eleven models had been delivered by V-J Day, in addition to one shipped to Wright Field for static test. The chronic airframe and engine headaches caused further Army evaluation of the ability of the C-69 to meet the Army's airlift needs. From May 1943, when the contracts still called for 260 airplanes, to the contract cancellations following V-J

Day (August 1945), C-69 production was progressively cut back. In May the Army's Production Engineering Section recommended cancellation of all C-69s, except for the first twenty, but was overruled. The C-69As were canceled but C-69 production was to continue, as was engineering on the C-69B. In June the contract was cut back to 210 airplanes, all C-69s, except the three C-69Bs. By April 1944 the C-69Bs were canceled as such and were to be delivered as plain C-69s. By March 7, 1945, the contract was further reduced to seventy-nine airplanes. In April, Lockheed proposed that the last fifty-three be built at a fixed unit price of $706,846, with the three redesignated C-69Bs at $2,632,728 apiece, quite a leap in cost.

In June 1945, it was decided that under an approved C-69 acceleration program the last C-69 would be delivered to the Army in February of the following year. Meanwhile, in order to obtain certification of the civil model, Lockheed would need a C-69 for that program. Thus it was agreed that Lockheed would complete the forty-eighth C-69 as a commercial airplane and then build an additional C-69 for the Army. It was later decided that the certification airplane would be the twenty-first C-69 instead of the forty-eighth. On June 22, 1945, the contract was cut further, to sixty-one C-69s and the three C-69Bs, with delivery to be completed by the next July.

The C-69 program got another setback on August 1, 1945, when Lieutenant General George decided on a production delay to complete engineering and tooling to put heavy floors and cargo doors on C-69s to make them as useful as the C-46s, C-47s, and C-54s already in service. This program might have been carried out except for another change in priorities. P-38 production was winding down and the B-17 program was finished. These events should have smoothed the way for the C-69, but another airplane got in the way. The new jet-propelled P-80 was now on the scene and AMC directed that it be given top priority, even if it meant cutting back on first, the C-69 program, and second, the remaining P-38 program.

This hectic prioritizing all became academic, however, soon after V-J Day,

when wartime military contracts were slashed or canceled industrywide. After considerable debate, the Army killed the last fifty C-69s on October 8, 1945, leaving only twenty-two of the originally planned 260.

Up to this time, the Army had accepted only fifteen C-69s, including the static-test article. Lockheed was able to buy the unfinished seven from the government for $402,500 each. Following its decision to continue production of the Model 49/049 for the airlines, Lockheed negotiated with the government to buy all government-owned C-69 tooling then in the factory.

After the final contract cancellations, most of the C-69s were sold back to Lockheed or to airlines through Lockheed. The few remaining in the Army inventory in 1947 were redesignated ZC-69, the Z prefix designating them as obsolete, an unusual status for airplanes of such recent manufacture.

The one C-69-5, Army serial number 42-94552, was static tested at Wright Field. Eleven of the fourteen flyable C-69s ended up as airliners. The first was used by Howard Hughes and then by Lockheed for testing, one became a depot for spares, and one crashed while still in the Army. The longest-lasting example, C-69-1, 42-94549, was registered to its last civil owner in July 1970, and was retired to the Pima Air Museum, Arizona, in 1975, twenty-seven years after it left military service.

C-69 Color and Markings

Except for the first C-69, which was camouflaged briefly, all C-69s were delivered in natural metal finish with the standard military insignia of the time, plus Army tail numbers.

Only one of the very few in postwar Army service is known to have carried the new buzz numbers, where each individual Army (later Air Force) airplane was identified by a two-letter code that designated its type and model, with the last three digits of the Army or Air Force serial number identifying the specific airplane of that type and model. For example, a C-69-5 was identified by CM-533, C for cargo, M for the C-69, and 533 the last three digits of the serial number 42-94533.

Postwar Model 049 Constellation

Following the post V-J Day cancellations of most of the remaining C-69 contracts, Lockheed made a major decision to keep the C-69 production line open, convert the on-hand and uncompleted C-69s as civil transports, and buy back surplus C-69s from the Army and refurbish them to civil standards. By this bold move Lockheed saved a lot of jobs and, since the 049/C-69 was pressurized, gained a two-year lead on Douglas' pressurized postwar airliner, the DC-6.

Certification and Early Operations

After V-J Day, one of the former Army C-69s was equipped to airline standards and presented to the Civil Aeronautics Administration (CAA, now the FAA) for certification testing. Approved Type Certificate (ATC) A-763 was awarded on October 14, 1945. Later, other short-body Constellations would be added to the same ATC, but new models with stretched fuselages and new wings required separate ATCs.

To obtain the necessary ATC, the C-69 had to undergo several modifications. For safety purposes, an engine fire detection and extinguishing system was incorporated because of the

Model 049-46 NC90825 on a Lockheed test flight before delivery to TWA on May 15, 1947. TWA leased it to Eastern, got it back, and retired it as a source of spare parts in 1964. TWA was the largest operator of the Model 049, with a total of forty-one. Of these, twenty-five were ordered new, seven were former USAAF C-69s, and nine were purchased from other airlines. NC86507, c/n 2028, was an 049-46-26 named Star of Madrid. It crashed on a training flight November 18, 1947. Lockheed

history of fire problems with the R-3350 engines. Further improvements were made to the cabin heating and cooling systems while more insulation was added to the cabin walls.

The interiors of the airplanes were completely redesigned to commercial standards from the bare-metal military interiors. Because the C-69 was now to be a civil transport, more windows were installed for passenger comfort and visibility as had been done on the C-69C-1. A major change in the cabin area was the installation of the galley in the front between the navigator's position and the passenger compartment. For international service, the navigator's area also contained bunks for off-duty crew members. For domestic service the area was changed to accommo-

Panair do Brasil, a Pan Am subsidiary, received its first 049-46-26 in March 1946, after it had been diverted from Pan Am. PP-PCF, c/n 2049, was originally registered NC88849. Panair named it Manoel de Borba Gato. Lockheed

Pan American had the second largest fleet of 049s, with twenty-two. This was increased when two ordered by Panagra were added to the fleet. NC88832, c/n 2032, was delivered February 19, 1946, and named Clipper Flora Temple. It was later transferred to Panagra and converted to Model 149 in 1955. Lockheed

date eight seats, which could be used by special request, for oversold flights, or as a lounge with card tables.

Even though the heating and cooling systems were improved, they were still totally inadequate. The installation of fans to help circulate the air had little effect. Because of this, cockpit crews used to call the flight attendants "Hot or Cold Running Hostesses," depending on their requests for temperature adjustments. Some areas would be too cold while others would be too hot; there just wasn't any in-between.

Whatever its shortcomings, the cabin of the 049 gave early postwar US domestic passengers their only taste of high-altitude pressurized comfort. TWA's prewar Boeing 307 Stratoliners had been drafted by the Army, but when returned were put back in service without pressurization. Pan Am did not operate its three 307s on domestic routes, and the pressurized Douglas DC-6 and Boeing 377 were still in the future.

Capital Airlines of Pennsylvania received its first converted C-69 in 1950 and went on to operate a total of twelve. An interesting feature on Capital's 049s was the forward Cloud Club Room lounge that could seat eight while the main cabin seated fifty-six. This is the third C-69, c/n 1963, serial 43-10311. It was leased to Pan Am, then became the oldest C-69 owned by an airline when Capital bought it in September 1950. Peter M. Bowers

American Overseas Airlines (AOA), a division of American Airlines, was the third airline to receive 049s and operated them from the United States to Scandinavia, Iceland, and Germany. AOA, along with its seven 049s, was purchased by Pan Am in September 1950. NC90922, c/n 2052, an 049-46-27, was initially named Flagship Denmark by AOA but became Clipper Mount Vernon when with Pan Am. Lockheed

BOAC was the first European operator to receive C-69/049s, with seven. Deliveries beginning in April 1946 enabled BOAC to offer the first service from London to New York by a European carrier, on July 1, 1946. Note that the letter G of the British registration on former C-69-6, serial 42-94557, now named Balmoral, had been covered over during Lockheed test flights. Lockheed

Dutch airlines KLM received its first 049 in May 1946 and soon started flights from Amsterdam to New York. Four 049-46-59s were purchased but were traded in for later models by 1950. PH-TAV 2069 ended up as a restaurant in Santiago, Chile, after a succession of owners. Lockheed

TWA's 049-46-26 N86516 Star of Ireland *in an air terminal setting typical of the 1950s. TWA leased it to Eastern for a year in 1956,* then sold it to Nevada Airmotive in 1962. Ed Peck

Further improvements included lockable doors on the two lavatories in the rear and a coat compartment. A storage bin was also located in the rear of the cabin and included emergency evacuation equipment. The original equipment included a Jacob's ladder, a flexible evacuation rope with solid rungs, to be attached to a key slot on the outside of the fuselage above the rear door. Passengers were expected to use the ladder or slide down ropes located at the cockpit door and the wing emergency exits. Both methods required a trapeze artist to negotiate, if the crew could even find the key slot, and were eventually replaced with emergency chutes. These were installed near the rear cabin door and the crew door in the cockpit and were a welcome change for passengers as well as crew.

Some of the modifications made for the C-69 were not needed on the 049, although they were retained. For some, these included the extra windows above the pilots' heads and the astro-

Air France operated four 049-46s between Paris and New York and other parts of the world. The airline was actually owned by the French government, which retained control of the airplanes until January 30, 1950. F-BAZB 2073 was delivered in June 1946. It was sold to TWA in February 1950 and became N9410H Star of London. *The airplane was scrapped in 1964.* Lockheed

dome in the navigator's area. They were eventually removed while the airplanes were in service. When the converted C-69s entered service, they still had the landing lights located in the nose instead of the wings. This location irked some pilots because if anything hit the nose it would break the cone, producing a very loud whistle until the airplane finally landed.

One area that should have been corrected was the original free-casting (nonsteerable) nose wheel. With the original unit, the only means of controlling the airplane on the ground was by using a combination of rudders and brakes. This worked fine until the airplane wasn't moving; then the rudders had no effect. The only solution was to lock one brake on the main landing gear box and, with a burst of power from the opposite engines, pivot on that set of wheels. This made for interesting situations when maneuvering in tight areas, and sometimes ground equipment would be blown around. The problem was finally corrected but not before sixty-two 049s, including the twenty-two C-69s, had been delivered.

By November 1945, Lockheed was able to announce orders for eighty-nine 049s from domestic operators TWA, Pan Am, American Overseas, Eastern, and Pan American Grace Airways (Panagra), plus foreign orders from KLM and Air France. (Not all of these orders were filled. Some airlines elected to take later models while others took 049s. Initial airline operators of 049s are listed in chapter 11.) Some domestic operators were forced to buy elsewhere because of a clause in Lockheed's contract with TWA. It stated that no airline could operate a Constellation in a west-to-east direction (meaning transcontinental) for a period of two years. This sent most competitors to different manufacturers while others, such as Pan Am and American Overseas, with their transatlantic routes, and Eastern with its north-south East Coast route, could order the 049 because of their route systems.

Lockheed and the airlines bought newly surplus C-69s from the government and Lockheed completed the unfinished C-69s still in the factory as civil airliners. Lockheed kept the C-69/049 production line open and turned out purely civil 049s into May 1947.

The first commercial operator of Connies was Pan Am, which flew one from New York to Bermuda on February 3, 1946. While Pan Am used a sizeable fleet of thirty-three Connies, of which twenty-two were 049s, it is TWA, the prime initiator of the design, that is most commonly associated with the Connie as its pioneering airline operator. TWA possessed by far the biggest Connie fleet—a total of 188, of which thirty-one were new or secondhand 049s. The fleet included a few secondhand acquisitions and encompassed four basic Connie models and three submodels. Not all were in service at the same time.

Capitalizing on the Connie's range, TWA started New York–Paris flights on February 6, 1946, followed by Los Angeles–New York service on March 1. Eastbound flights took nine hours, fifteen minutes with one stop, easily beating competing airlines that were then using Douglas DC-4s. Westbound flights took eleven hours, twenty-five minutes. Some publicity flights with reduced payload made it coast-to-coast nonstop, but such regular service had to wait until 1953. The efficiency of the Connie on the North Atlantic run is reflected by the fact that TWA had competition from five other airlines that were operating Connies on that route by 1947.

While the 049 as delivered did not come up to the 1940 contract specifications because of its austere military changes, it was still far ahead of the competing Douglas DC-4 in speed, range, passenger capacity, and the now all-important cabin pressurization. These capabilities forced Douglas to redesign and stretch the DC-4 into the DC-6, with 2,400hp Pratt & Whitney R-2800 engines in place of the 1,450hp R-2000s, greater passenger capacity, and pressurization.

Three different engines were offered to the airlines in the 049. The 2,200hp Wright 745C18BA-1, the 2,200hp Pratt & Whitney R-2800-26C14G, which was similar to the R2800-83 then being tested in the Army's XC-69E, and the very similar 2,300hp British Bristol Centaurus. However, all customers stayed with the Wright in spite of its continuing problems.

The 049s started with a gross weight of 86,250lb, but were soon allowed higher weights as a result of various structural reinforcements and landing gear improvements.

These allowed increases were reflected by suffix letters added to the 049 designation as follows:

Model No.	Gross Weight (lb)
49	86,250
49A	90,000
49B, C	93,000
49D	96,000
49E	98,000

The figures are taken from current FAA specification sheets which, as with other FAA and Lockheed paperwork,

Linea Aeropostal Venezolana (LAV) received the first of its two 049-46s on October 31, 1946, and started service between Caracas and New York on March 21, 1947.

The registration YV-C-AMI was transferred to an LAV 1049G-82 Constellation in 1947, long after c/n 2082 was sold to Braniff in August 1955. Lockheed

identify the airplanes as 49s instead of 049s.

It should be noted that these suffix letters are not used to reflect basic model improvements, such as Model 749 to 749A, and so on. With hindsight, it might have been more logical to use dash numbers to identify the different weights of the basic 049 model.

Nearly half of the 049s were converted to 049Ds. Seven became 049Es, but at that weight were underpowered. Other pertinent specification and performance figures for the original 049 Connie are: span 123ft, 0in; length 95ft, 2in (or 98ft, 2in with radar); empty weight 55,345lb; maximum speed 329mph; cruising speed 275mph; service ceiling 25,500ft; range 2,290 statute miles with 18,400lb payload and 3,680 miles with 7,800lb payload.

Flyaway factory prices ranged from $685,000 to $720,000, but some of the surplus C-69s had been bought for as little as $100,000. After their replacement in first-class service by later Connies, TWA's 049 fleet was converted to high-density all-coach seating in 1955.

It should be noted that the Connies built as civil airplanes after the twenty-two C-69s started a new serial number series at 2023, the twenty-three indicating the twenty-third 049 airframe. These serial numbers carried on to 2088, the eighty-eighth and final 049, delivered May 5, 1947.

Early Problems

Integration of the Constellation into the postwar civil airliner fleet was not without some serious problems. The first accident occurred on a Pan Am flight from New York to Europe on June 18, 1946, when an uncontrollable fire broke out in the number four engine soon after takeoff. Eventually the engine dropped off the airplane, taking the fire with it. A successful belly-landing was made in a grass field without injury to passengers or crew. At first thought to have been an induction fire, it was later discovered to be a failure of the driveshaft for the cabin pressurization pump. The shaft had apparently broken off and, acting like a large steel machete, proceeded to slice off everything connected to the back of the engine, including the fuel and hydraulic pumps. The immediate fix was to install a pipe-like housing over the driveshaft to keep it from flailing around if it became disconnected.

Even though the passengers and crew had been in immense danger, there is a lighter side to this story. After the airplane had bellied in, the passengers were seated on the ground on the airline's blue blankets and were served tea and coffee along with sandwiches and cakes. It seems that there was a flight attendant onboard who would not let anything, not even a brush with death, ruin cabin service.

Another serious accident occurred about three weeks later, on July 11, when a TWA 049 had an in-flight fire near Reading, Pennsylvania. The smell of burning insulation was noticeable in the cockpit. The external hatch was opened to air out the cockpit, but this only proceeded to feed the fire. The Connie crashed in a field with only one person surviving out of the six aboard. The fire was traced to a through bolt, the electrical connection between the generators and the main electrical bus as it entered the cabin in the forward baggage area. This was also the area where the hydraulic reservoirs were located and when hydraulic fluid leaked onto an overheated or arcing through bolt an insulation fire could result.

As a result of this incident, the entire Constellation fleet was grounded to fix this particular problem and others that were giving trouble. In-flight fires with the carbureted R-3350 engines in Army B-29s resulted in Wright replacing carburetors with fuel-injection systems. Modifications made to the 745C18BA-1 engines in Connies to incorporate fuel injection resulted in an engine designation change to BA-3. With the modified engines and other changes, the Connies were back in the air by August 23. The rescinded ATC A-763 was reissued on October 14, with the constructor's numbers of later 049s added to it. Once the grounding order was lifted, both TWA and Pan Am began to operate Constellation flights again on domestic and international flights.

Of the eighty-eight 049s and C-69s built, eighty-two went to the following first-owner airlines:

Airline	Quantity
TWA	31
Pan Am	20
AOA	7
BOAC	6
KLM	6
Intercontinental US	4
Air France	4
Capital	2
LAV (Lineas Aereas Venezolanas)	2

It should be noted that these airlines are "first-tier" users of these airplanes, even though the former C-69s were not new. When an airliner was passed to another airline, that line became the "second-tier" operator of a used airplane. Some airlines, like TWA, operated 049 Connies obtained from other airlines as well as its own factory-new 049s, so were therefore both first-tier and second-tier users of the Model 049.

TWA's Star of Ireland. Ed Peck

Refining the Constellation—
Models 649 and 749

Even though the Model 049 Constellation proved to be an exceptional airliner, Lockheed realized from the beginning that it was only an interim design until a true commercial version could be produced. In May 1946, work began on a revised and updated version in response to an Eastern Air Lines request.

Model 149

Although it did not appear until after the Model 749, the Model 149 is presented first because of its numerical sequence.

No Connies were built under this designation. A 1940 proposal for a lightweight civil Connie with 875hp Wright GR-1820-G3 engines was not followed up. After the war the designation was used to identify 049s that had been fitted with Model 749 outboard wing panels containing 565gal fuel tanks to increase the total fuel capacity from the original 4,690 to 5,829gal.

As the Model 149-46, the 049 conversion with 2,400hp 745C18BA-3 engines was added to the original 049 ATC on March 18, 1948.

Model 649

Starting with the basic 049, Lockheed increased the available power by installing 2,500hp 749C18BD engines and improving the cabin heating, air-conditioning, and ventilation systems. Integral fuel capacity remained at 4,690gal while the maximum takeoff weight was increased to 94,000lb—4,000lb more than the 049A. The dimensions remained the same.

Other improvements included revised engine cowlings, new propellers with reversible pitch, and an increase in flap deflection. To reduce cabin noise and vibration, Lockheed installed shock-mounted wall panels comprised of several layers of fiberglass insulation, air spaces, and fire-resistant fabric. The interior skin was rubber-mounted from the exterior, thus

No Model 149 Constellations were built as such. All were Model 049s fitted with 749 outboard wing tanks for increased range. N88838, c/n 2038, was an 049-46-26 delivered to Pan Am on January 1, 1946. It later went to Panair do Brasil in December 1953. Peter M. Bowers

ensuring a reduction in noise and vibration, since only the cabin windows had direct contact with the structure. Another new feature was the addition of small wide-angle-lens windows in the cockpit to enable the flight engineer to see the engines.

The 649 represented a 50 percent redesign of the basic Constellation, offering a higher cruising speed, greater economy, and vastly improved passenger accommodations. This model was considered the first true civil Constellation, and more like the original dedicated airliner of 1940 than a refurbished Army cargo plane. Dubbed the Gold Plate Constellation by Eastern, it was first flown on October 19, 1946.

Initially many airlines, including Eastern, TWA, KNILM, KLM, and Air France, ordered this new version.

In an attempt to attract operators that preferred the Pratt & Whitney R-2800 engine over the relatively untried Wright engines, Lockheed offered the R-2800 as an alternate engine in the 649. The R-2800s had actually been tested for the Army on the XC-69E, but in the end only R-3350s ever powered production reciprocating-engine Constellations. Another proposed version of the Constellation was to be built in England with Bristol Centaurus engines. The Ministry of Supply in London had concluded that a license-built Constellation could be built in about the same time as a new design. This version became known as Project Y while another version, Project X, involved fitting Lockheed-built Constellations with the Bristol engines. Both projects were eventually canceled because they were too costly.

The Speedpak

An innovative feature, first offered on the last twelve 049s, and available on all subsequent standard Constellations, was the Speedpak. This special under-fuselage freight container, developed at the request of Eastern and tested on an Army C-69, provided an additional 395cu-ft of cargo space. This was enough for 8,300lb of cargo and allowed operators to take advantage of maximum payloads on short or medium routes (up to 1,000 miles). Having been both wind-tunnel and flight tested, the Speedpak produced no adverse effects on handling and reduced the speed by only 12mph. Loading and unloading were conducted with a self-contained electric hoist that lowered the Speedpak to the ground. This, it was claimed, could be completed in two minutes, thus shortening ground time for handling cargo and baggage. Ground handling of the Speedpak was facilitated by a pair of semi-recessed wheels mounted at both ends which enabled it to be rolled away from the aircraft.

TWA had planned to buy eighteen 649s, but an airline strike in October 1947 forced the airline to cancel the order. This left Eastern as the only carrier to receive the 649 Constella-

Eastern Air Lines was the only customer for the Model 649. It received its first 649-79-12 on March 19, 1947, and had it in first-class service in June. NC101A, c/n 2518, was actually Eastern's first 649, but was retained by Lockheed for testing under an NX license until October 10, 1947. It was later converted to Model 749A. Lockheed

tion. Eastern had originally ordered fourteen 049s in September 1945, but switched to the superior 649 and also placed an order for fourteen Speedpaks.

As with the 049s that had been built as airliners instead of C-69s, the Model 649 was in a new serial number series. This series also included the contemporary Model 749, which actu-ally started the new series at 2503. With the 649, airplane deliveries got out of sequence with model numbers. Eight 749s were delivered before 649s, and both the 649 and 749 received their ATC approvals on the same day.

The 649 first flew on October 18, 1946, and as 649-79 was added to the 049 ATC on March 14, 1947. Eastern Air Lines received the second of its fourteen-plane order (c/ns 2518–2524, and 2529–2535) on May 13, 1947. The first article was retained by Lockheed until October. Eastern paid an average price of $850,000 each for its 649s.

Model 649A

Some of the modifications devel-oped for Models 749 and 749A could be retrofitted to the 649, resulting in the

The instrument panel of the Model 649 was much neater than that of the 049. Improvements included installation of the propeller-reversing controls on top of the four throttles, and locating the autopilot at the bottom of the control pedestal. Lockheed

These two photos show just how much cargo the Speedpak could really hold. Practically all of the items in the fore-ground could fit in; there was no tieing-down of the cargo. The cables and hooks for attaching to the airplane are visible on the right side of the loaded Speedpak photo. Lockheed

649A with fuselage and inner wing reinforcements as well as modified brakes. These enabled the maximum takeoff weight to be increased to 98,000lb. Only six 649As were built as such, and were delivered to Chicago & Southern Airlines with 2,400hp 749C18BD-1 engines as 649A-79. These were added to the 049 ATC on December 20, 1949. With modifications, Model 749s could be upgraded to 749As. Eastern had its 649s upgraded to 649A standards before they were updated further to 749s in 1950.

Model 749

In the face of stiff competition from the newly introduced Douglas DC-6, Lockheed decided to produce an improved version of the 649. Many of the carriers that had ordered the 649 wanted an airliner with increased range. The 749 was the first Connie to have the additional 1,555gal outer wing tanks built on the production line. Oddly, the first eight 749s of sixty delivered were built before the 649s, and started a new c/n series at 2503. The increase in gross weight to 102,000lb brought a new requirement for fuel dump valves to reduce weight quickly in case of engine failure during a maximum-weight takeoff. For the longest range flights, passenger seating was reduced because of the need for quarters for additional onboard relief crew. Long-distance flights also made sleeping berths desirable, so that feature, as originally planned in 1940 and offered on, but not adopted for the 049, finally appeared on some 749s.

Since the fuel tanks were located in the outer wings, the maximum takeoff weight could be increased without modifying the airframe. This was possible because of the fuel distribution in the wing and also the sequence in which it was used.

A new type of exhaust system was fitted to the last few 749s built, this being the "jet stack" exhaust. It was a form of exhaust thrust augmentation that added a claimed 15mph to the maximum cruising speed. Earlier 749s could be retrofitted with the system but cabin noise increased substantially. The 749 carried the same payload as the 649, but had 1,000 miles greater range.

This Pacific Northern 749A-79-33, serial N10401, c/n 2661, was PH-TFG on lease from KLM in 1956; hence the KLM rudder stripes. This view shows the length of the stowed Speedpak relative to the length of the fuselage. Peter M. Bowers

Chicago and Southern Airlines (C & S) was the only customer for the Model 649A. It received the first of six, N86521 2642, on August 1, 1950. Passenger accommodation ranged from forty-nine to fifty-seven. C & S merged with Delta Air Lines in 1953. Lockheed

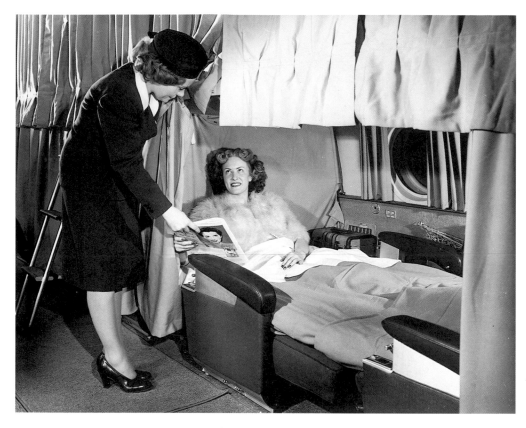

The lower berth of a Model 649, made up. Ladder behind cabin attendant provided access to the upper berth, which was fitted with a smaller round window. Lockheed

Most of the airlines that had ordered 649s switched their orders to the superior 749. Lockheed retained one 749 for itself; the other fifty-nine went to the following first-tier owners:

Airline	Quantity
KLM	13
TWA	12
Air France	9
Eastern	7
Aerlinte Eireann Teoranta*	5
Pan Am	4
QANTAS	4
Air India	3
LAV	2

*An airline based in Ireland.

An oddity of serialization appeared with the Eastern 749s. Their c/ns, 2610, 2611, and 2614–2618, appeared in a new block of numbers that had been started by the first 749A at 2600.

After the initial postwar boom, production rates of the Constellation dropped off drastically. Some airlines cut back on the number of Connies they had ordered while others canceled them altogether. Only seventeen airplanes were scheduled for delivery in all

The Model 749 was a greatly improved version of the C-69/049 that was developed specifically as an airliner. This is KLM's 749-79-33 PH-TFD on a factory test flight before delivery on May 31, 1950. Lockheed

of 1948, and Lockheed seriously considered closing down the production line. Fortunately, the armed forces stepped in with a dozen 749 orders, two for the US Navy and ten for the new US Air Force that had just replaced the US Army Air Forces. This was just enough of a boost to keep the line open until civil orders picked up with the Model 749A.

Model 749A

Further improvements to the 749 resulted in a gross weight of 107,000lb. The first twelve built were for the military, ten ordered by the Air Force as C-121A and two by the Navy as PO-1W. These started a new numbering series at 2600. The first civil 749A, c/n 2619, still with BD-1 engines, was added to the 049 ATC on February 15, 1949.

The first of fifty-nine civil models went to Air India on October 20 of that year. Altogether, thirty-two 749As were delivered to US airlines while twenty-seven went to foreign airlines. The

TWA received its first Model 749 in March 1948 and operated twelve of the type along with twenty-eight Model 749As, three of which were purchased sec- ondhand. By early 1951 a record number of fifty-eight scheduled Constellation flights were operated across the Atlantic weekly. ATP/Airliners America, Ken Bader

Australia's QANTAS Airways received all four of its 749-79-31 Constellations in October 1947. VH-EAB, c/n 2565, has six sleeping berth pairs on each side of the cabin as shown by the small upper berth windows, not seen on a Constellation since the first two C-69s. Lockheed

749A was also the last of the short-fuselage Connies to be built.

It is interesting to note that Lockheed offered a civil freighter version of the 749 that was similar to the Air Force C-121. This was an unpressurized model with a 6ft high by 7ft, 8in wide cargo door installed at the rear of the main cargo compartment and another smaller door installed just behind the cockpit. The modification also included a watertight cabin, removable and stowable compartments, skylight windows, and cargo tie-down rails and fittings. A small lavatory was installed just behind the cockpit and operators had a choice of having all or part of the cabin refrigerated. Up to 3,433cu-ft of cargo space was available. Even though this version offered potential operators outstanding performance and load-carrying capabilities, none were produced. However, many other 649 and 749 Connies retired from passenger service were converted to freighters and ended their careers as such. Still others were converted to spray planes.

The fifty-nine civil 749As were delivered to the following first-tier airline customers:

Airline	Quantity
TWA	26
Air France	10
KLM	7
Chicago & Southern	6
Air India	4
South African	4
Avianca*	2

*Aerovias Nacionales de Colombia, a Colombian airline.

The 649 and 749 proved to be the most reliable of all the Constellations built. These final short-fuselage Connies were extremely popular with pilots because of their exhilarating acceleration and excellent go-around capabilities. A total of 132 749 and 749A Constellations were built, including twelve military versions, before production ended in 1951. Production was discontinued so that a newly developed version, the Super Constellation, could be produced. (See chapter 8 for details.)

Because of its popularity, reliability, and good economics, the 749 outlasted later models to become the last American-operated Connie to carry passengers in scheduled airline service. This was a Western Airlines flight from Juneau to Anchorage, Alaska, on November 26, 1968. The airplane was a fifth-hand 749-29-22 that Western had acquired with five others when it absorbed Pacific Northern Airlines in February 1968.

ZS-DBR, c/n 2623, was the first of four Model 749A-79-50s delivered to South African Airways starting in April 1950. Day-passenger capacity on the London-Johannesburg route was fifty-three. The sleeper version had eight berths and twenty-six sleeper seats. Many airlines had their name in their own language on one side of the airplane and in English on the opposite side. Lockheed

Military Model 749 Constellations

At the low point in postwar airliner sales, Lockheed was losing money on the Constellation and was seriously thinking of shutting down the production line. Fortunately, the Connie received orders from the new independent US Air Force—which had been created from the former US Army Air Forces on September 18, 1947—and the US Navy. The two orders, placed in late 1947 and early 1948, totaled twelve airplanes, ten for the USAF and two for the Navy. They helped keep the Constellation line open and the planes' performance brought additional and much larger orders (see chapter 9).

The USAF Constellations continued to be called Constellations, but the Navy models were named Warning Star in a combination of their specialized naval mission and Lockheed's practice of naming its airplanes after celestial bodies.

Air Force 749A Constellations

When the USAF ordered its postwar Constellations as C-121 in February 1948, it was under conditions far different from those that complicated the history of the original C-69. This time, instead of a new design that pushed the state of the art under the handicap of low wartime priorities, a troublesome engine, and material shortages, the USAF bought a minor variant of a well-proven and in-production airliner, which it designated C-121A. The C-121A was what the C-69B was to have been—a reliable, long-range, pressurized transport for either passenger or cargo use. However, the new model was not an Air Force sales triumph for Lockheed. Major competitor Douglas had equaled the Connie with its postwar seventy-four-passenger DC-6A, which featured 2,500hp R-2800 engines, reinforced floor, and a cargo door. In view of

this, the USAF ordered 101 DC-6As as C-118As, while ordering only ten C-121As.

Why the C-121 designation instead of a higher series C-69 for such a similar airframe? Although the Model 749 Connie was structurally and aerodynamically similar to the earlier 049 and 649 with the same R-3350 engine, and the new military version was not significantly different from the "heavy" but unbuilt C-69B, the USAF chose to call its second-generation Connie the C-121 rather than a C-69F or something similar. Why?

It couldn't have been for structural or equipment differences, which were far less than those between, say, a B-17D and a B-17F. It couldn't have

been because of the change of service name; many other designs that started on US Army contracts continued under USAF contracts. It couldn't be because a design went out of military production and then was revived, although that gap might have had something to do with it. Perhaps it was just that the C-69 had been such a headache that the Air Force wanted to forget it and make a new start with a new model number.

(*Note*: Although delivered after the C-121B, the C-121As are described first for the convenience of alphabetical progression.)

C-121A

The ten-plane, $11,425,000 C-121A order (no block numbers) was placed

The fifth C-121A, 48-613, was the first to be fitted with APS-10 weather radar (note nose radome). Carrying a special paint scheme, it was assigned as VC-121A to Gen. Douglas MacArthur. *Photo taken at Pusan, Korea, in October 1950.* Norman E. Taylor

57

Still with its special paint but stripped of its V designation, 48-613 served the Special Air Missions organization until 1966. It was then assigned to NASA and was re-tired to the US Army Museum at Fort Rucker, Alabama, in March 1970. Gordon S. Williams

The second C-121A, 48-610, when being used by President Eisenhower as VC-121A Columbine II. *It remained the presidential airplane until November 1954. Retired in* 1969, it was sold for parts but has now been restored to its former glory (see chapter 16). Gordon S. Williams

on February 10, 1948, but only the last nine, serial number 48-609/617, c/ns 2601–2609, were delivered as C-121A cargo planes. These were designed for heavy cargo and alternate personnel loads, either forty-four seated passengers or twenty litter cases and their medical attendants. The floor was reinforced and a single upward-swinging cargo door, 112x72in, was installed on the left side toward the rear of the cabin. Engines were civil 2,500hp 749C18BDs and gross weight was 107,000lb.

The first C-121A was delivered on December 31, 1948, and the last on March 24 of the following year. All were soon assigned to special passenger work as VC-121As. Shortly before retirement in the late 1960s, they reverted to the standard C-121A designation. Most were eventually sold for conversion to spray planes, but one went to Ethiopian Air Lines via Lockheed, and 48-613, after use by the National Aeronautics and Space Administration (NASA), was retired to the US Army Aviation Museum at Fort Rucker, Alabama.

VC-121A

After routine operations with the Military Air Transport Service (MATS), all nine C-121As were sent back to the factory in 1950 for modification as VC-121As, with deluxe interior furnishings and additional cabin windows. They were then assigned to the 1254th Air Transport Squadron at Washington, DC, for use as VIP transports. For this work the prefix V, in use from 1945 to date, was added.

C-121 Operations

Seven of the C-121As were used in support of the 1948-1949 Berlin airlift, not by flying into Berlin, but by flying more than 5.9 million miles over the North Atlantic in their first month on that assignment.

As VC-121As, the USAF short-fuselage Connies carried some very notable passengers in an interesting situation where the US military commanders actually had larger and slightly faster VIP transports than the president himself. Gen. Dwight D. Eisenhower used VC-121A 48-614, which he named *Columbine*, as his personal transport when he was commander of Supreme Headquarters Allied Powers Europe (SHAPE) from December 1950 to June 1952. (The columbine is the state flower of Colorado, his wife's native state.) Later, when he was president, Eisenhower had VC-121A 48-610 for a while, which he named *Columbine II*. After its retirement from the USAF, *Columbine II* was sold to a civil operator who used it as a source of spare parts for other C-121As he had bought. In 1990 it was back in the air, restored as *Columbine II*. (See chapter 16 for more details.)

Gen. Douglas MacArthur used VC-121A 48-613, named *Bataan*, during the Korean campaign. Later identified as SCAP, for Supreme Commander Allied Pacific, it was eventually acquired by NASA for use at the Goddard Space Flight Center in Greenbelt, Maryland.

The first article on the ten-plane C-121A order was rushed to early completion as the one-only C-121B, 48-608. Photographed on a factory test flight in November 1948, before special markings and VC designation were applied. Lockheed

After being redesignated VC-121B, 48-608 served the brass in Washington, D.C., until *1966. Note the later Boeing VC-137B jet in the background and distorted proportions* *of national insignia on the fuselage.* John T. Wible

The first US Navy PO-1W, a Model 749A airframe equipped with special electronics *as an early-warning patrol plane. Compare cabin window pattern with earlier* *C-69 and with Navy R7V-1 in chapter 9.* Lockheed

After withdrawal from their VIP roles, the VC-121As were downgraded and reverted to basic C-121As, possibly with the questionable PC-121A designation. Six VC-121As, serials 48-610–614 and 48-617, became VC-121Bs. By 1969 all of the C-121As had been declared surplus and were sent to Davis-Monthan Air Force Base in Phoenix, Arizona, for storage. All soon found civil buyers and most went on to long careers as sprayers and cargo planes.

C-121B

The first of ten C-121As ordered, serial 48-608, c/n 2600, did not become a cargo plane. As C-121B it was rushed to completion ahead of the others with a VIP interior in the belief that it would become the presidential airplane of Thomas E. Dewey, who was expected to defeat incumbent President Harry S. Truman in the 1948 presidential election. Truman won, however, and elected to retain his single Douglas VC-118, serial 46-505. The C-121B was to have been named *Dewdrop* if used by Dewey. The name was later applied to another USAF airplane.

The C-121B was used by such high officials as the secretary of defense and the secretary of the Air Force. Naturally, in view of its missions for such big brass, it was redesignated VC-121B.

VC-121B

As a VIP transport, 48-608 had accommodation for twenty-four day passengers and fourteen night or sleeper passengers, with ten berths and four sleeper seats, and, of course, no cargo door or heavy floor.

Serial number 48-608 was retired in December 1969, after reverting to C-121B. It was sold to Aircraft Specialties of Mesa, Arizona, for conversion to a sprayer, and received the civil registration N608AS (note the combination of the Air Force serial number and the new owner's initials).

Six other VC-121As, 48-610–614 and 48-617, were upgraded to VC-121B standards and were redesignated as such. They, too, were downgraded to C-121B and found new owners after military retirement.

US Navy 749A Constellations

The US Navy began its Connie operations with a radar-equipped 749A variant that was initially designated PO-1W; *P* for patrol, *O* for Lockheed in Navy designations since 1930, *-1* for the initial configuration, and the suffix letter *W* indicating the special mission of the airplane as early warning. The two letters made it natural to call the planes "PO Ones." The two

In 1952 the Navy changed the designation of the two PO-1Ws to WV-1. This is the second one, photographed in Hawaii soon after the change. A two-star admiral's placard was placed on the nose between the door and the figure 10. David W. Lucabaugh collection

Directional stability problems resulting from the large radomes made it necessary to enlarge the vertical tail surfaces of the two PO-1Ws, but only the outer fins were enlarged as can be seen in this photo of the first PO-1W. Lockheed

The last C-121A, downgraded from VC status, photographed in August 1967. The top of the fuselage was white and prefix

letter O preceded tail number 80617. By this time, USAF airplanes used as VIP transports carried the designation "United

States of America" on the fuselage instead of "U.S. Air Force." David W. Lucabaugh via Norman E. Taylor

The first Model 749-79-22 was delivered to Mexican airline Aerovias Guest S.A. in 1947, went to Pan Am, and then to Air France in 1949. It was later taken into the

French air force and used as a research test bed with both military and civil markings. Alain Pelletier

PO-1s, Navy Bureau of Aeronautics numbers (BuNos) 124437 and 124338, c/ns 2612 and 2613, were ordered to prove the feasibility of the Airborne Early Warning (AEW) concept.

PO-1W

The PO-1W, powered with the same 2,500hp civil engines as the C-121A and also fitted with a left-side cargo door, was easily distinguishable by a large radome fitted above the fuselage that resembled the conning tower of a submarine. This housed the height-finding radar while a smaller radome mounted below the fuselage contained the surveillance radar. Eight small stublike antennae protruded from the top of the fuselage and two more were located on the right side. Another feature of the PO-1W was the installation of weather radar in the nose.

A crew of thirty-one was carried, including the pilots, radar operators, and technicians. The cabin interior was taken up by radar monitoring stations and other equipment, with galley facilities and rest bunks to provide a degree of crew comfort.

The first PO-1W flew on June 9, 1949. To counter directional instability problems caused by the radomes, the outboard fins and rudders were enlarged as would later be done on the Model 1049 Connie.

WV-1

The Navy changed the designated mission from *P* to *W* for early warning in 1952, eliminated the old *O* for Lockheed in favor of *V* for Vega, and redesignated the airplanes as WV-1. Later, the radomes and special equipment were removed and the airplanes were trans-

ferred to the FAA in 1958 and 1959, where they received the civil registrations N119 and 120. Both were later transferred to the USAF, which operated them with civil registrations N1192 and 1206 in areas where overflights by airplanes with USAF radio call signs would not be appreciated. N1192 was put in storage in 1966 and eventually scrapped. N1206 was sold in 1966 and had a succession of civil owners that did not fly it.

Military 749A Constellation Colors and Markings

The military 749A Connies were delivered in natural metal finish with the appropriate military markings and insignia. In 1953, both services began painting the upper fuselage of passenger-carriers in white as a means of reducing solar heating of the cabins. Also, both services began to use variously sized areas of Day-Glo fluorescent orange on noses, rear fuselages, and tails to increase visibility of the aircraft in the 1950s and 1960s. This paint was different than the Insignia Red and later Red-Orange used for special Arctic and Antarctic markings.

During the service of the C-121, the USAF revised the tail serial number by adding the prefix zero (not the letter *O*) followed by a dash to the number already on the tail to indicate an airplane that was then over ten years old. The purpose of this was to avoid duplication of tail numbers by a plane with a ten-year-old number, as for example C-121A 48-612 (applied as 8612) and the later 58-612 (also applied as 8612) on a Lockheed T-33A.

Super Constellations

Even with all the successes of the Constellation in airline use, it was recognized almost immediately that an enlarged version would be needed to keep pace with the increasing demand for air travel. Lockheed realized that a stretched version of the Constellation would ensure reduced costs per seat per mile and also allow lower fares over existing types. During tests an updated version of the 749 was flown at an all-up weight 30,000lb greater than that of the 749A. This demonstrated the degree of stretchability in the basic design, and research began in 1949.

Rather than build a new prototype from scratch, Lockheed bought the first C-69, which had become the XC-69E, from Howard Hughes, who had bought it after the war. Lockheed changed the model number to 1049 and changed the c/n from the original 1961 to 1961-S, S for stretch. The registration number was changed from the original pre-Army NX25600, also used by Hughes after the war, to N67900. (In 1948 the second letter in

A Lockheed family reunion—1951. In the foreground is the second production Super Constellation. Next, an Air France 749A-79-46 delivered in September 1951, and three pre-World War II Lockheeds, the Model 10 Electra (1934), the Model 12 Electra Junior (1936), and the Model 18 Lodestar (1939). In the background is one of the two giant XR6O-1 Constitutions (Model 89) built for the US Navy in late 1946. Lockheed

The prototype of the Super Constellation series was the first C-69 that became the XC-69E and was bought by Howard Hughes after the war. Lockheed bought it from Hughes, reinstalled Wright R-3350 engines, and stretched the fuselage. The old round windows, pilots' windshield, and the original small vertical tail surfaces were retained. Lockheed

The first Super Constellation, Model 1049-53-67, flew on July 14, 1951. After certification testing, it was delivered to Eastern Air Lines in March 1952. New features included rectangular cabin windows, revised cockpit windows and engine nacelles, and enlarged tail surfaces. Lockheed

US civil aircraft registrations NC, NX, and so on, was deleted. Thus NX25600 became N25600.)

Model 1049

The Model 1049, soon named Super Constellation, incorporated an 18ft, 4³/₄in stretch of the standard Constellation fuselage for a total length of 113ft, 7in. This was accomplished by inserting two equal-length "plugs" in the fuselage ahead of and behind the wing. New uprated 2,700hp Wright engines were installed to offset the increased weight of the fuselage. Along with the increase in power, the fuel capacity of the basic 1049 was increased to 6,550gal. This was made possible by the installation of a 730gal center-section fuel tank in the same Model 749A wing, which increased the airplane's range some 500 miles over the 749A. Larger rectangular windows, nineteen on the right side of the fuselage and sixteen on the left, replaced the smaller circular windows of the previous models and provided 85 percent greater visibility for the passengers.

A new seven-pane windshield, 3½in higher than before, provided an additional 7in of headroom in the cockpit while also increasing the pilots' view by 21 percent. The area of the vertical tail was increased to stabilize the extra length of the fuselage by splicing in a straight section on the previously fully elliptical fins and rudders at the level of the horizontal tail as had been done on the PO-1Ws. The rudders were now metal covered. To give passengers increased comfort, the cabin heating and cooling systems were substantially improved and the pressurization system could now provide a 5,000ft cabin altitude at 20,000ft.

To increase structural strength and minimize weight of the Super Constellation, Lockheed pioneered new manufacturing methods. One method featured newly developed integrally stiffened skin panels—large wing panels milled out of single pieces of metal. This not only saved weight but also eliminated the need for numerous panels and rivets to hold the wing together.

Because of its increased weight and bulk, Lockheed had intended to use the new turbo-compound version of the R-3350 engine in the Model 1049 from the start, but the engine was not available when the airplane went into production. Noncompound 956C18CA engines of only 2,700hp at takeoff while running on 115/145 octane gas were installed and resulted in an airplane that was slower than the Douglas DC-6B which had been developed to compete with earlier Connies. As a result, sales of the basic Model 1049 were not good, with only twenty-four being built.

Overall, a total of 550 design improvements went into the 1049 and Lockheed was able to certify it at a maximum gross weight of 120,000lb—4,000lb more than was guaranteed. The prototype made its initial flight on October 13, 1950, and started a new numbering series at 4001. First flight of a production 1049 was on July 14,

The Model 1049 Super Constellation was a stretched version of the Model 749 with larger vertical tail surfaces and square rather than round cabin windows. N6901C was the first Model 1049-45-67 for TWA and was delivered August 6, 1952. Lockheed

The increase of seven inches of pilot headroom is immediately apparent here because the upper console is now out of the picture. Also, the autopilot controls are on the bottom of the center console, with individual controls for the elevator, rudder, and ailerons. Lockheed

1951. As Model 1049-53, it received a new ATC, 6A-5, on November 28 of that year.

(Since the issuance of ATC A-763, the method of issuing ATCs was changed. Instead of coming in consecutive order from a single FAA office, ATCs were now awarded from several separate FAA regions throughout the country. Thus ATC 6A5 identified the fifth airplane certificated in FAA Region Six.)

The first airline to order this new version was Eastern, which signed for ten 1049-53-67 Super Constellations on April 20, 1950, but later increased the order to fourteen. Even though an extra fuel tank could be carried in the center section, Eastern's airplanes were not so equipped. However, one could be installed at a later date. Pan Am leased one of Eastern's 1049s from June until November 1955, to test the suitability of the Super Constellation for its routes, but Pan Am did not order any. Eastern started commercial service on December 7, 1951. Used only on domestic service, Eastern's 1049s had a three-person cockpit crew and two cabin attendants.

TWA ordered ten Model 1049-54-80 Super Constellations on December 5, 1950, and the first was delivered in May 1952. Like Eastern, TWA ordered its 1049s without the center-section fuel tank. Once the airplanes were in service, TWA requested some modifications to increase their speed. These included extending and improving the engine nacelles, extending the propeller spinners rearward, closing off one of the three wing intakes for the

Close-up of the tail of Colombian Avianca 1049E-55, c/n 4554, shows off the larger vertical tail surfaces of the Super Constellation created by adding a small straight section to the original ellipse just above and below the horizontal stabilizer. Victor D. Seely collection

cabin cooling, and removing the wing walkway paint. These modifications increased the airplane's speed by as much as 12mph at 20,000ft.

TWA inaugurated nonstop Los Angeles—New York flights on October 19, 1953. These took a little over eight hours and brought up problems with the pilots' union. Westbound flights had to stop at Chicago for a crew change until the pilots agreed to a ten-hour working day for the transcontinental run. Although these new models proved to be a valuable asset to TWA, they were underpowered and were the first Connies to be retired and sold by the airline.

On June 9, 1953, the Model 1049-53 was deleted from ATC 6A5. All had been converted to 1049-54s with 2,800hp 975C18B-1 engines, and the new approval was issued under 6A5 on May 14, 1952.

Models 1049A and 1049B

The 1049A and 1049B were military versions of the Super Constellations. Model 1049A applied to US Navy WV-2 and WV-3 and USAF RC-121D models. Model 1049B applied to Navy R7V-1 transports, USAF RC-121Cs, and the single VC-121E presidential *Columbine III*.

Model 1049C

Model 1049C was the first civil 1049 to attain its intended performance goals, thanks to use of the turbo-compound engines that finally became available for civil use in 1953. The turbo-compound derived its name from the three power-recovery turbines on each engine. Each turbine was fed from the exhaust of six cylinders and fed its power back to the engine through a fluid coupling. This turbine input increased the engine's power by 20 percent.

The 872TC18DA-1 delivered 3,250hp for takeoff on 115/145 octane fuel. This allowed a gross weight of 133,000lb and upped the cruising speed to nearly 300mph. The 1049C put Lockheed ahead of Douglas again and beat the new DC-7, which also used the Wright turbo-compound engine, into the air by three months. However, the turbo-compound engines were quite troublesome, and frequent in-flight failures resulted in 1049Cs and subsequent 1049s earning the derisive nickname of "The World's Best Trimotor."

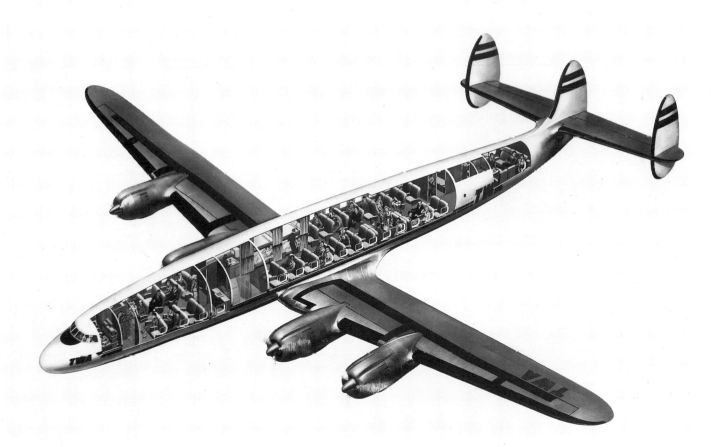

Interior arrangement of a TWA Super Constellation. The cabin compartmentation was adopted for the longer cabin of the Model 1049. *Other airlines had their own variations.* Lockheed

Eastern Air Lines was the first customer for the Super Constellation. Model 1049-53-67, NC62026, c/n 4002, is the second of fourteen used on the lucrative New York–Miami route. Eastern also bought sixteen 1049Cs and ten 1049Gs, and was one of the last airlines to retire the type. Lockheed

TWA was the only other customer for the basic Model 1049, with ten. This is its second 1049-54-80, N6902C 4016, delivered in May 1952. It was lost in a collision with a DC-7 over the Grand Canyon on June 30, 1956. Lockheed

The most serious problem was excessive exhaust flaming and afterburning out of the turbine hoods of the three power-recovery turbines. When climbing at night, the airplane's exhaust flames reached all the way back to the trailing edge of the wing. This not only had adverse effects on the structural strength of the wing, it also made passengers more than a little bit nervous. Even after continued research and development, the problems continued throughout the models, including the Super-G. In 1953, after $2 million, nine months of intense research, and numerous flight tests, a cure was finally found. It involved installing a 2in wide ring of $1/2$in armor around each turbine, plus several other detail changes.

With the additional power, inner wing reinforcements, and changed fuselage structure, Lockheed not only increased the maximum takeoff weight but also raised the cruising speed to 330mph. However, these increases were made at the expense of range, which decreased by 300 miles. Additional improvements included a new main landing gear retraction system, oil transfer and fuel management systems, and relocated passenger doors. Additional insulation and more modern soundproofing made the 1049C one of the quietest airliners of its day.

The first flight of the 1049C, which started still another c/n series at 4501, was on February 17, 1953. As 1049C-53, it was added to the basic 1049 ATC on June 9 of that year. Forty-eight 1049Cs were built, with deliveries to seven airlines beginning the same month. Passenger arrangements varied, with long-range overseas versions reducing passenger capacity for additional fuel and rearranging the forward part of the cabin for additional crew quarters.

When improved 988TC18EA-3 or EA-6 engines were installed, 1049Cs became 1049C-03 and -05. Some 1049Cs were later converted to 1049Es and a few of those were further modified to 1049Gs.

The seven airline customers for the Model 1049C and the quantities delivered to them were:

Airline	Quantity
Eastern	16
Air France	10
KLM	9
Trans-Canada	5
QANTAS	3
Pakistan-International	3
Air India	2

Model 1049D

The four 1049Ds were essentially civil versions of the Navy R7V-1 transports (Model 1049B) with rectan-

The 1049 was steadily improved. This is 1049C-55-83 N6225C, for Eastern Air Lines, with turbo-compound engines. Intervening A and B variants were military models. Lockheed

One of the power-recovery turbines of the turbo-compound Wright R-3350 engines can be seen in this view. Each engine has three turbines, which feed the energy of exhaust gases back to the engine for a 20 percent power increase. Lockheed via San Diego Aerospace Museum

gular windows and commercial 972TC18DA-1 engines. They were built near the end of R7V-1 production and just before the start of USAF C-121C production. The 1049D was the world's largest commercial cargo transport when it appeared with a gross weight of 133,000lb and capacity for 38,750lb of payload. Freight doors were installed with the forward door measuring 5ft, 1½in wide. Smaller personnel doors were installed within the cargo doors. A new heavy-duty freight floor made up of magnesium planks was installed and incorporated tie-down rings, seat attachments, and stretcher fittings for casualty evacuation. The interior could be fitted to seat up to 109 passengers with quick convertibility being a feature of the design. Although offering tremendous load-carrying capabilities, only four, c/ns 4153–4166, registrations N6501C–6504C, were built for Seaboard & Western, with delivery in August and September 1954. The Model 1049D-55 was added to the basic 1049 ATC on August 12 of that year.

Model 1049E

The twenty-eight 1049Es were passenger planes similar to the 1049Cs but had structural beef-up that allowed the gross weight of the 1049D. The 1049E-55 was added to the 1049 ATC on May 26, 1954. Some 1049Es were later converted to 1049Gs. A few airlines switched their

Close-up shows the great differences in engine cowling between the Model 1049 Constellation with turbo-compound engines and the earlier nonturbo models. Lockheed via San Diego Aerospace Museum

Pakistan International Airlines bought three of the forty-eight Model 1049Cs. This is its third, 1049C-55, AP-AFS, c/n 4522, delivered in April 1954. The airline was founded in 1951, and later added two Model 1049Hs. Lockheed

1049C orders to 1049E, along with additional 1049E orders from other airlines.

The following eight airlines were first-tier customers for the twenty-eight Model 1049Es:

Airline	Quantity
QANTAS	9
KLM	4
Avianca	3
Iberia*	3
Trans-Canada	3
Air India	3
LAV	2
Cubana	1

*Iberia Lincas Aereas Espanolas, a Spanish airline.

Soon after the 1049E was ATCed an event took place on July 15, 1954, that was to number the days of Connie production. Boeing flew its jet-powered Model 367-80, prototype of the 707 transport and the harbinger of an entirely new era of air transport. The British De Havilland Comet had opened the jet transport era in 1952, but did not have the range or the capacity to pose a serious threat to the latest piston-engine types. The 707 and its derivatives and copies were far ahead of the Comet and would soon completely displace the Connies and DC-7s from the world's major airline routes.

Model 1049F

Model 1049Fs were the civil versions of the USAF C-121C passenger-cargo transports which were similar to the US Navy R7V-1 (Model 1049B).

Model 1049G

Once the 1049Cs and Es were in service, it became obvious immediately that increased range was needed. In response, Lockheed developed the 1049G by installing 609gal wing-tip fuel tanks. These increased the fuel capacity to 7,750gal and the airplane's range by 700 miles. Even though Wright did not have an engine with increased horsepower, they did have an improved turbo-compound which enabled the maximum takeoff weight to be increased to 137,500lb. Some were later modified to allow a gross weight of 140,000lb. Much of this allowed increase came as the result of structural revisions for the same size wing, which

Seaboard & Western was the only customer for the Model 1049D freighter, which featured reinforced flooring and a cargo door fore and aft. This is the fourth 1049D-55, N6504C, c/n 4166, delivered in September 1954. The 1049D had a partially extended Fowler flap, apparently to slow the Connie down for formation flying with a slower camera plane. Lockheed

There were twenty-eight Model 1049Es built. This is the first of two 1049E-55s for the Venezuelan airline LAV, YV-C-AMS, c/n 4561, delivered in December 1954. Lockheed

Lufthansa began its initial postwar overseas service on June 8, 1955, with eight 1049G-82s. This is D-ALAK, c/n 4602, the first, delivered April 1, 1955. The 609gal wing-tip tanks, added to the G-model, increased the range by 700 miles. Lockheed

Wing-tip fuel tanks were an option on Model 1040Gs, not standard equipment. This is the first of five 1049G-82s for Air India, VT-DIL 4646, delivered June 11, 1956. Service to Europe usually had the cabins arranged for sixty-three passengers. Lockheed

featured machined and integrally stiffened skins.

The most notable external feature of the 1049C was the addition of 609gal wing-tip tanks as introduced on the Navy WV-2 and the USAF RC-121D. Altogether the Super-G introduced 107 improvements over the 1049E, including new propeller hubs and chord-wise deicer boots on the wings and tail. Cabin soundproofing was improved by installing an inner wall made of sound-insulating plyboard and blankets of a new fiberglass material to the inner surface of the cabin skin. Rubber shock pads on the engine mounts also helped to reduce noise and vibration. Weather radar could be installed in the nose of the airplane; this increased the total length to 116ft, 2in.

The 1049G first flew on December 7, 1954, and as 1049G-82 with 92TC18DA-3 engines giving 3,250hp was added to the 1049 ATC on January 14, 1955. Service with Northwest Airlines started on July of that year. TWA called its 1049Gs Super-Gs and painted the designation on the tail.

Altogether, 102 1049Gs were built, forty-two for US airlines and fifty-nine for foreign lines. Range was notably improved over the basic 1049, at 4,140 miles, with an 18,300lb payload and 5,250 miles with 8,500lb. Even so, the 1049G did not have dependable non-stop New York–Paris range.

With 102 built, the Super-G was the major production version of the civilian Model 1049. Beside the Hughes Tool Co., which bought one, sixteen airlines were first-tier customers for Model 1049G as follows:

Airline	Quantity
TWA	28
Air France	14
Eastern	10
Lufthansa	8
KLM	6
Varig*	6
Air India	5
Northwest	4
Trans-Canada	4
Cubana	3
TAP**	3
Thai Airways International	3
Iberia	2
LAV	2
QANTAS	2
Avianca	1

Slick Airways bought three of the fifty-nine convertible cargo-passenger 1049H models. This is the first of Slick's three, N6937C 4830, delivered September 17, 1959, very near the end of Model 1049 production. Slick later leased another six 1049Hs before going out of business at the end of 1965. Lockheed

*Empresa de Viacao Aerea Rio Grandense, a Brazilian airline.
**Transportes Aereas Portugueses, a Portuguese airline.

Model 1049H

The last of the Super Constellation series to be developed was the 1049H. Lockheed took a hard look at the possibility of producing a multipurpose airliner capable of quick conversion between passenger and cargo roles. With the 1049H, Lockheed produced a convertible model of the Super-G incorporating the cargo modifications of the 1049D. In a matter of hours a 1049H could be converted from a strictly cargo plane to a passenger airliner. This conversion was carried out by installing toilets, interior lining panels, cabin luggage and baggage racks, and seating for up to ninety-four passengers. A buffet or bar could also be installed depending on the operator's requirements. Gross weight and performance were the same as for the 1049G. The first flight of the 1049H was on September 20, 1956, and as 1049H-82 it was added to the 1049 ATC on October 9.

A total of fifty-nine H-models were built, all going to the fifteen following airlines and one leasing organization:

Airline	Quantity
Flying Tigers	13
California Eastern	5
Gulf-Eastern	5
Seaboard & Western	5

Airline	Quantity
National	4
REAL*	4
TWA	4
Air Finance Corp.	3
KLM	3
Slick	3
Pakistan International	2
QANTAS	2
Resort	2
Trans-Canada	2
Dollar	1
Transocean	1

*Redes Estadvais Aereas Limitada

Just after the first 1049Hs had been completed, Wright produced an improved version of the turbo-compound engine rated at 3,400hp. This enabled Lockheed to increase the maximum takeoff weight to 140,000lb. The 1049H Super Constellation was then capable of transporting 40,203lb of freight under 5 percent overload condition.

When production of the Super Constellation series ended, a total of 259 had been produced for civil operators. Flying Tigers acquired the last production Super Constellation in November 1958, and the last undelivered Super Constellation was purchased by Slick Airways on September 30, 1959. Quickly replaced on the trunk passenger routes by the Boeing 707 and Douglas DC-8 jets, the 1049Hs remained with the major airlines as freighters, although many were passed down to the all-freight lines.

Military Super Constellations

A gathering of military Super Constellations on August 14, 1955, with a few civil models mixed in. In the foreground is President Eisenhower's VC-121E Columbine III. Behind it on the left is a Navy R7V-1 transport. Beyond the three civil models, and distinguished by their high dorsal radomes, are (left) a USAF RC-121D and (right) a Navy WV-2. Beyond the WV-2 is a Navy R7V-2 with turboprop engines. Lockheed

An early Navy R7V-1 on a factory test flight. Note the nose radome for weather radar and the use of fewer and smaller cabin windows on a fuselage longer than that of the earlier C-69. Lockheed

The Navy's PO-1W/WV-1s were so successful that the Navy ordered additional AEW aircraft developed from the newer Model 1049 Super Constellation. Since the certification trials for the 1049 were completed, Lockheed decided to use the prototype, referred to as "Old 1961," for future military testing. This airplane went through many test phases for the numerous radomes to determine the shapes and their aerodynamic performances. The resulting airplane was the PO-2W Warning Star, later redesignated WV-2. The PO-2W designation was never used.

While the Navy initially ordered 244 WV-2s, it also ordered similar airframes configured as cargo-transports under the designation R7V-1. Because of their simpler installations, the R7V-1s were delivered before the first WV-2s.

The USAF also procured Super Constellations under continuing C-121 designations. Considering the great differences in aircraft size, weight, and power, it is hard to understand why the Navy distinguished between its Model 749A and 1049A Connies with only a dash number change. It would have seemed logical to identify the 1049A as a W2V-1, a new model. Lockheed did regard it as a new model. The USAF, on the other hand, had precedent for using only series letter changes to cover major airframe changes; for example, the change from the Boeing B-17D to B-17E.

For convenient reference, the military Super Connies are grouped first by as-ordered Navy models, then USAF, and finally by the Combined Services designations adopted in 1962. The models for each service are listed alphabetically by model designation rather than by sequence of production and delivery. To keep separate lists complete, there is some duplication in the USAF and Combined Services lists.

US Navy Super Constellations

It should be noted here that the Navy had been procuring Lockheed airplanes under two manufacturer's letters—*O* for the main Burbank plant as used since the XRO-1 of 1931 through the PO-1W of 1948, and *V* for the products of the subsidiary Vega plant since 1942. In 1952 the Navy gave the products of both plants the same letter *V*.

R7V-1

The initial PO-2W and WV-2 orders were followed by Navy orders for fifty standard cargo-transport equivalents designated R7V-1 (Lockheed Model 1049B-55-75). Navy BuNos were 128434–128444, 131621–131629, 131632–131649, 131651–131659, and 140311–140313. Engines were 3,250hp R-3350-91s from USAF stock. Accommodation was for seventy-two troops, but the cabin windows were limited to eight round portholes to a side. The R7V-1s started a new Lockheed c/n series at 4101.

Because of their lack of special features other than a rear cargo door and reinforced floor, the first R7V-1s were delivered before the first WV-2s, starting in 1942. To divide up the work of the Military Air Transport Service (MATS), the Navy transferred thirty-two R7V-1s to the USAF, where they became C-121Gs. In September 1962, the Navy's remaining R7V-1s were redesignated C-121Js.

R7V-1P

A single R7V-1, BuNo 131624, was fitted with cameras for aerial survey work in Antarctica, so received the special-purpose suffix letter *P* to indicate a photoplane version of a standard model. Operated by Navy Squadron VXE-6, the R7V-1P was named *Phoenix 6* while operating in Antarctica in 1959.

R7V-2

To service-test turboprop engines in large airplanes, four of the R7V-1s under construction were completed as R7V-2s (BuNos 131630, 131631, 131660, and 131661) with 6,000 shaft horsepower (shp) Pratt & Whitney YT34-P-12B engines. The last two were

transferred to the USAF as YC-121Fs. The first flight was on September 1, 1954.

During Navy tests, the R7V-2s were put through many "extreme emergency" procedures to see how the engines would perform. On one flight an R7V-2 reached a speed of 479mph while in a dive (comparable to the maximum diving speed of a P-38 Lightning). Another R7V-2 took off at a weight of 166,400lb, or roughly twice the maximum gross weight of the C-69s.

One R7V-2, 133661, was loaned back to Lockheed and then sent to Rohr Aircraft in Chula Vista, California, for the test installation of an Allison Model 501 turboprop engine in the number four position to develop the installation for the new Lockheed Model 188 Electra II civil transport then being designed. While carrying its Allison engine, 133661 was nicknamed *Elation*, a combination of Electra and Constellation, before being returned to standard turboprop con-

The single R7V-1P was a photoplane conversion of an R7V-1. Named *Phoenix 6*, it was used by Squadron VXE-6 in Antarctica in 1959. The cabin top was painted dark gray as on WV-2 patrol planes, not for camouflage, but to absorb solar heat for cabin warming. US Navy via Harold Andrews

Close-up of the three-blade propellers on the first Navy R7V-2. Very wide blades were required to absorb the engines' 6,000hp without increasing the diameter of the propellers. Lockheed

figuration and transferred to the USAF as the second YC-121F.

WV-2

Out of a planned purchase of 244 Model 1049A-55-70s, the first few were ordered as PO-2Ws, but by the time deliveries began in 1954 the designation had been changed to WV-2. After diversion of some to other Navy designations and the transfer of seventy-two to the USAF only 142 Navy Super Connies were delivered as WV-2s. BuNos were 126512, 126513, 128323–128326, 131387–131392, 135746–135761, 137887–137890, 141289–141333, 143184–143230, and 145824–145841. These started still another Lockheed c/n series at 4301.

The WV-2 had a new feature for military Connies, 609gal wing-tip tanks, a detail shared with the turbo-prop R7V-2s and some civilian Model 1049Gs. The upper radome, containing APS-45 height-finding equipment, was the same size as on the PO-1W and WV-1, but the lower radome, containing APS-20B sea-search radar, was larger than previously.

The WV-2 had an endurance of thirty hours with a crew of twenty-six and nearly six tons of radar equipment. To ease off-duty fatigue, sleeping bunks and a galley were provided. These enabled crew members to rest or prepare hot meals and drinks during flights. While in service, a number of WV-2s were converted to WV-2Q standards with sophisticated electronic countermeasures (ECM) and direction finding.

In 1962 the WV-2s were redesignated EC-121Ks.

WV-2E

Another area that Lockheed studied was the rotating-dish radar that could enhance the capabilities of current AEW radar. Taking the first WV-2, 126512, from the production line, Lockheed installed the new 37ft AN/APS-70 aerial surveillance radar integrally into a 40ft rotating dish called a rotodome, thus producing the WV-2E. The entire dish assembly weighed nine tons and was mounted on a large pylon on top of the fuselage, which still contained the APS-45 radar of the WV-2. The normal ventral AN/APS-20 radar of the WV-2 was removed along with its radome, since it was not needed. The radar dish slowly rotated while the airplane was in flight and enabled the WV-2E to detect targets at up to three times the range of the APS-45 radar. The WV-2E, intended to be the prototype for a planned W2V-1 to be powered with Allison T-56A turbo-props, made its first flight on August 6, 1956.

The W2V-1 program was canceled after the competing Grumman W2F-1, with a nonrotating radome above the fuselage, won the competition. The WV-2E retained its separate identity, was used for further research and development, and became the EC-121L in 1962.

WV-2Q

A small number of WV-2s, including BuNos 131390–131392, 135753, and 143209, were modified for ECM missions. They were redesignated EC-121Ms in 1962.

WV-3

Eight WV-2s, BuNos 137891–137898, were completed as weather reconnaissance planes. These had a

After the first C-69 Constellation was stretched to become the prototype of the Super Constellation, it was fitted with dummy radomes and used again as the prototype of the Navy WV-2 early-warning patrol plane. The vertical tail surfaces have now been enlarged to Super Constellation size. *Lockheed*

A WV-2 patrol plane in then-standard overall Seaplane Gray coloring over a radar picket ship off the coast of Newfoundland on March 14, 1957. *US Navy via Steve Kraus*

crew of twenty-six, including stations for aerographers and aerologists, plus special air-sampling equipment. Because of their special equipment and use, the WV-3s earned the nicknames "Storm Seeker" and "Hurricane Hunter." One WV-2, BuNo 141323, was modified after delivery as a ninth WV-3. In 1962 the WV-3s were redesignated WC-121Ns.

The Navy received its first WV-2s in October 1955, but it was nearly three years before they were in full squadron service. AEW missions were flown over the Atlantic and Pacific oceans by squadrons VW-2, -3, -11, and -13. Land-based radar eliminated the need for such patrols, so the last AEW mission was flown by a VW-11 EC-121K on August 26, 1965. WV-3/WC-121Ns continued to service the weather reconnaissance role into 1976.

US Air Force
Super Constellations

The first Super Connies for the USAF were transfers from the Navy contract for R7V-1s, and became special-purpose radar-equipped RC-121Cs. In the following description, however, the USAF designations are listed first by the basic model designation, C-121C, then by the special-purpose variants such as EC-121C and JC-121C, then NC-121D, RC-121D, and so on. Designations above C-121D were for specially built models or 1962 redesignations of existing models.

C-121C

In spite of its basic designation, the C-121C was the USAF's third Model 1049 Connie (1049F-55-96), flying some three years after the RC-121C and two years after the RC-121D. Thirty-three, serials 54-151–183, with continuing c/ns from the Navy R7V-1, were ordered. One additional C-121C, 51-3840, was acquired by conversion of an RC-121C/TC-121C.

In general detail, the C-121C followed the Navy R7V-1 Super Connies, with 3,510hp R-3350-54 (Navy) engines, 135,000lb gross weight, and twenty airline-type rectangular passenger windows to a side instead of the R7V's eight round windows. Both models shared the Model 1049D detail of two cargo doors on the left side of the cabin, one ahead of and one behind the

wing. The forward door was slightly smaller than the rear.

During their USAF careers many C-121Cs were redesignated when used for other than straight transport-cargo duties. Those designations included EC-121C, JC-121C, RC-121C, TC-121C, and VC-121C. Many were transferred in their later years to Air National Guard (ANG) units and remained in service into the late 1970s.

EC-121C(1)

Confusion results from the near-simultaneous use of the prefix *E* with two separate meanings. From 1946

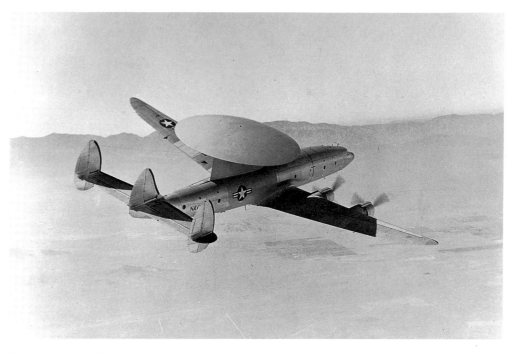

The single WV-2E was the first WV-2 diverted to a program to evaluate the AN/APS-70 surveillance radar in a rotating dish-shaped antenna called a rotodome.

The system was not adopted for Constellations, but the WV-2E retained its unique feature after becoming the EC-121L in 1962. Lockheed

The first of eight WV-2s were modified as WV-3 weather reconnaissance planes. They had additional antenna and radome, and a large rubber deicer boot on

the leading edge of the dorsal radome. Color is overall glossy light gray. US Navy via Harold Andrews

into the 1960s the letter *E* was used to mean exempt, that is, the airplane was in a special test status that exempted it from most routine technical order compliance whose accomplishment would delay the test program. Some E-planes were involved in USAF test work, but others were used by private organizations on airplanes supplied by the Air Force on bailment contracts.

One C-121C, 54-160, became such an EC-121C. It was fitted with an exterior, cupola-like test station for two people, under the fuselage and just behind the wing. Those crew members were involved in special visual-acquisition testing. The test observer would try to detect targets visually while an evaluator checked his or her performance against radar and infrared acquisitions of the same predetermined targets.

EC-121C(2)

In the redesignations of 1962, the *R* for reconnaissance (search) prefix was replaced on the RC-121Cs by the prefix *E* for special electronics installations. The RC-121Cs (also the TC-121C trainers) then became EC-121Cs.

JC-121C

The *J* prefix identified a special test status from which the airplane could be returned to its original configuration with relative ease. This was an alternative to putting anything other than a true prototype airplane on experimental status as indicated by the prefix *X*. Two C-121Cs (54-160 and 54-178) and one TC-121C (51-3841) were so modified for electronic systems testing and research as JC-121Cs.

RC-121C

It is hard to understand why the passenger-cargo C-121C and the radar-equipped RC-121C Warning Star should have shared the same series letter when they were such notably different aircraft. The ten RC-121Cs (Model 1049B-55-84, 51-3836–3845) were AEW types with very few round cabin windows and owed much to the earlier Navy R7V-1 and the contemporary WV-2 for their structure and aircraft systems. In fact, the USAF models were diverted from a Navy WV-2 order but did not have the WV-2's wing-tip fuel tanks.

The RC-121Cs entered service with the USAF's Air Defense Command (ADC) in October 1953. They were later redesignated TC-121C and EC-121C.

TC-121C

Nine of the RC-121Cs were redesignated TC-121Cs to indicate their use as trainers, but were redesignated EC-121Cs in 1962. One, serialed 51-3840, was converted to a basic C-121C transport but whether with proper C-121C rectangular windows is not known.

VC-121C

Four C-121Cs, serials 54-167, -168, -181, and -182, were given VIP interiors and served with the Special Air Missions squadron based in Washington, DC.

This EC-121C, 54-160, was used to check visual detection of ground targets against radar detection of the same targets. A visual observation station was located under the fuselage behind the wing. Gordon S. Williams

The prefix letter J identified a special test status for an otherwise standard airplane. This C-121C, 54-178, was photographed at Wright-Patterson AFB in Dayton, Ohio, on May 21, 1967, when fitted with a test radome on top of the fuselage to become a JC-121C. Peter M. Bowers

The prefix letter J when used with similar suffix letters did not mean the planes looked alike. This former RC-121C, 51-3841, was temporarily a TC-121C before being fitted with a topside cupola for visual detection of missiles reentering the atmosphere under the designation of JC-121C. Douglas D. Olson

EC-121D

EC-121D is the 1962 redesignation of the RC-121D.

NC-121D

The status prefix *N* identifies airplanes that have been modified so extensively for test purposes that they cannot be conveniently returned to their original configurations. The single NC-121D, 56-6956, was such a plane.

It was a former Navy WV-2, 143226, transferred to the USAF for a program to measure radiation from the reentry of high-speed bodies into the earth's atmosphere. The program was called TRAP III, for Terminal Radiation Airborne Program. The NC-121D, stripped of its WV-2 radomes and anti-submarine warfare (ASW) equipment, was sent to Lockheed Aircraft Service Co. for the installation of an array of optical and electronic sensors on top of the fuselage and two visual sighting stations on the forward fuselage that were essentially enlarged navigator's astrodomes.

RC-121D

The major (and last) production Model 1049 Connie ordered for the USAF was the RC-121D, Lockheed Model 1049A-55-86. Seventy-two were ordered over a three-year period: 52-3411–3425, 53-533–556, 53-3398–3403, 54-2304–2308, and 55-118–139. One additional RC-121D, 54-183, was obtained by modifying a C-121C. Its major outward difference from the RC-121C was the addition of WV-2 type radomes and wing-tip fuel tanks.

In the redesignations of 1962, the RC-121Ds became EC-121Ds, with some becoming EC-121Js.

VC-121E

The single VC-121E, 53-7885, was something of an oddity. It was started as a Navy R7V-1, BuNo 131650, but was modified on the production line as a special USAF VIP airplane (Lockheed Model 1049B-35-97) with rectangular windows replacing the R7V-1 port-holes. It became President Eisenhower's *Columbine III* and served on into the Kennedy administration until

This specially marked plane of the National Guard Bureau, 54-181, is one of four C-121Cs fitted with plush interiors for the carriage of VIPs under the designation of VC-121C. Norman E. Taylor

In 1962 the Air Force RC-121Ds were redesignated EC-121Ds. This one, 52-3423, has been stripped of its upper radome. Note the overall dull-gray finish. Norman E. Taylor

US Air Force EC-121Rs were former Navy WV-2/EC-121Ks stripped of their radomes, camouflaged, and used for specialized electronic missions in Vietnam. P. B. Lewis via David W. Menard

Final production model Super Constellation for the USAF was the RC-121D, distinguishable from the similar radome-equipped RC-121C by the addition of wing- tip fuel tanks. Outwardly similar to Navy WV-2s, the RC-121Ds became EC-121Ds or EC-121Js in 1962. Paul Pauson via Norman E. Taylor

replaced by a Boeing VC-137C jet in October 1962.

Columbine III's special equipment featured a teletype for transmitting and receiving classified messages, a television set, and an air-to-ground telephone link. After every 1,000 flying hours, the airplane was returned to Lockheed for a virtual rebuild to ensure proper operation of every part on the airplane.

After being retired as the presidential aircraft, called "Air Force One" only when the president was aboard, the VC-121E became just another VIP transport assigned to the 89th Military Airlift Group at Andrews AFB, Maryland. Since 1980 it has been in the USAF Museum at Wright-Patterson AFB in Dayton, Ohio.

YC-121F

To evaluate turboprop engines in large transport-cargo airplanes, the Navy ordered four R7V-1s to be completed as R7V-2s (Lockheed Model 1249) with 6,000shp Pratt & Whitney

The single USAF VC-121E, 53-7885, started down the Lockheed production line as Navy R7V-1, BuNo 131650, but was completed as the plush Columbine III for President Eisenhower. Note that the more numerous cabin windows are rectangular. Lockheed

Built as a WV-2, BuNo 141292 has had many subsequent designations. In 1962 it became an EC-121K, then the NC-121K shown here, and ended up as an EC-121P. K. Buchanan via David W. Menard

The weight of US Army airplanes was limited by agreement with the US Air Force, but an exception was made when this former WV-2/EC-121K was borrowed from the Navy, modified, and used as the JC-121K under its original Navy BuNo. J. Sherlock via David W. Menard

Two of the Navy's turboprop R7V-2s, 131660 and 131661, were transferred to the USAF where they became YC-121Fs 53-8157 and -8158. This first one, in USAF markings, still has its Navy BuNo 131660 on the tail. Longer nacelles were necessary to move its lighter turbine engines forward to maintain airplane balance. Lockheed

T34-P-6 turbine engines. Two of these were transferred to the USAF as service-test YC-121Fs, serial numbers 53-8157 and -8158. Except for retaining the R7V's eight round windows, the YC-121Fs were equipped to C-121C standards.

The use of turboprop engines was to evaluate the engines, not to improve the performance of the airframe in which they were installed. The USAF simultaneously evaluated turboprop engines in two reengined Boeing C-97s as YC-97J and in two Convair C-131s as YC-131C.

C-121G

The Navy transferred thirty-two of its fifty-one R7V-1s to the USAF, which assigned the designation C-121G and a single block of USAF serial numbers, 54-4048–4079. The original BuNos were random, not a single block of consecutive numbers. Except for the absence of rectangular cabin windows, the C-121Gs were similar to C-121Cs. Four became TC-121Gs, with one later being redesignated VC-121G.

TC-121G

Four C-121Gs, 54-4050–4052, and 54-4058 were modified as crew trainers designated TC-121Gs. One later became the single VC-121G with plush interior for VIP passengers.

Joint Services Constellations—1962 and On

In September 1962, the aircraft designating systems of the USAF and Navy were combined. This had relatively little effect on existing USAF designations but caused abandonment of many long-standing Navy designations. Under the new system, Navy Connies had their *R* and *W* type designations changed to various series of C-121, starting with C-121J. In the pre-1962 Navy system, special-purpose modifications of equipment were identified by a suffix letter, as WV-2Q. After the change, the special purpose was identified by a prefix, with the WV-2Q becoming the EC-121N.

In 1962 the WV-2s were redesignated EC-121Ks with no change of equipment or coloring. This one was photographed near Naval Air Station (NAS) Argentia, Newfoundland, in April 1964. US Navy

The second YC-121F, 53-8158, used on scheduled transport runs by the Continental Division of the Military Air Transport Service (MATS). US Air Force

After 1962, some Air Force C-121s modified for new missions were also redesignated with higher series letters while others reflected the 1962 standardization of special-purpose prefixes, as *R* to *E*.

EC-121C This model represented the 1962 redesignation of USAF RC-121Cs and TC-121Cs.

EC-121D EC-121D was the 1962 redesignation of USAF RC-121D models.

EC-121H Redesignation of forty-two USAF EC-121Ds upgraded in 1962 to use special electronic equipment, including a computer to feed research data directly to the Semi-Automatic Ground Environment (SAGE) system of the North American Air Defense Command (NORAD).

C-121J The 1962 redesignation of Navy R7V-1s.

EC-121J Two Air Force EC-121Ds, 52-3416 and 53-137, upgraded with additional electronics. This particular redesignation resulted in the Navy and the USAF both having C-121 Connies with the same series letter.

NC-121J Four R7V-1/C-121Js were converted to NC-121Js during the Vietnam War for use as airborne television and radio studios and transmitters. The conversion installed extendable antennas above and beneath the fuselage. The NC-121Js arrived in Vietnam in 1967 and remained until 1970.

EC-121K Model represents the 1962 redesignation of Navy WV-2s. One became a YEC-121K and several others became NC-121Ks.

JC-121K Surprising US Army use of EC-121K, BuNo 143196, for electronic systems testing. The Army Connie, outwardly resembling the USAF JC-121C, was a far heavier airplane than the types allowed Army aviation by USAF-Army agreement.

Because of this weight limitation, the JC-121K did not become Army property. It was borrowed from the Navy and flew with its Navy serial

The single JC-121K was Navy WV-2 143196 transferred to the US Army and modified similarly to the JC-121C for missile tracking. Although carrying Army markings, the white-painted JC-121K operated under its original Navy bureau number. Duane Kasulka via Norman E. Taylor

After Navy WV-2 141292 became an EC-121K in 1962, it was diverted to a test program as NC-121K as shown here, and later became an EC-121P. It was equipped with additional radomes and antennae, and modified wing-tip tanks. C. Eddy via Norman E. Taylor

number but was marked US Army. Its purpose was to observe and track test firings of Army ground-launched missiles. To accomplish this, the former WV-2 was stripped of its radomes and Navy search equipment. An unpressurized cupola was built on top of the fuselage to house advanced optical and infrared tracking equipment installed behind optical glass windows on the left side of the cupola. Additional camera windows were installed on the left side of the fuselage, and the rear cargo door was modified so it could be opened in flight.

The JC-121K was returned to the Navy after its Army tests but was not restored to WV-2/EC-121K configuration. Instead, the Navy declared it surplus and put it in storage at Davis-Monthan AFB in January 1969.

NC-121K Former Navy EC-121Ks and the YEC-121K with extensive modifications for special missions. The single designation NC-121K does not mean that all airplanes with that designation were identical.

ENC-121K One NC-121K, BuNo 141292, placed on exempt status.

YEC-121K One NC-121K, BuNo 128324, used to service-test new equipment, with the Y for service-test added to the airplane designation.

EC-121L The Navy WV-2E, BuNo 126512, redesignated.

EC-121M Navy WV-2Qs redesignated.

WC-121N Navy WV-3s, with new W for weather prefixes.

EC-121P Several Navy EC-121Ks were fitted with upgraded submarine detection equipment and redesignated. Three were transferred to the USAF as JEC-121Ps.

JEC-121P Three EC-121Ps, BuNos 143189, 143199, and 143200, transferred to the USAF from the Navy for systems avionics testing. They retained their original BuNos instead of getting new Air Force serials.

EC-121Q Redesignation of Air Force EC-121D/EC-121H serial numbers 53-541, -556, 55-120 and -128 following installation of upgraded electronics for electronic reconnaissance and electronic countermeasures missions.

EC-121R Thirty former Navy EC-121Ks and EC-121Ps transferred to the USAF for use in Vietnam. They all got new USAF serial numbers in a single block, 67-21471–21500, although the original BuNos ranged from 137895 through 143294.

These models were stripped of their radomes and AEW equipment for use as airborne relay stations for trans-

This camouflaged EC-121R is one of thirty former Navy EC-121Ks and P-models turned over to the Air Force. Stripped of their radomes, they were reequipped with special electronic equipment to relay seismic intrusion data on enemy activity in Vietnam. Douglas D. Olson

The WV-2Qs were redesignated EC-121Ms in 1962. This is 145940 photographed landing at the US naval air station in Atsugi, Japan, in 1974. H. Nagakubo via Jim Sullivan

The twenty-two USAF EC-121Ts were redesignations of fifteen EC-121Ds and seven EC-121Hs. This one, 53-548, has overall glossy light gray finish and the dorsal radome removed. Photographed at McClellan AFB, Sacramento, California, in March 1973. Norman E. Taylor

The military services of only two countries besides the United States—Indonesia and India—used Super Constellations, but none were first-tier customers. This 1049G-82 was formerly VT-DJX of Air India and received Indian air force serial number BG579 before being transferred to the Indian navy. Norman E. Taylor

missions from the Air-Delivered Seismic Intrusion Devices (ADSID) dropped behind enemy lines to detect clandestine personnel movement and to direct air attacks against them. The EC-121Rs were the only Connies to be camouflaged since the original C-69 of 1943.

EC-121S Five USAF C-121Cs, 54-155, -159, -164, -170, and -173, were converted to the standards of the EC-121Q and were used by the Tactical Electronic Warfare Group of the Pennsylvania ANG in the late 1960s. These could be distinguished from the standard C-121Cs by a profusion of antennas and radomes.

EC-121T USAF electronic reconnaissance conversions of fifteen EC-121Ds and seven EC-121Hs. The EC-121Ts were operated by Air Force Reserve (AFRes) squadrons until their retirement in 1968.

Air National Guard Super Constellations

After reaching the end of their useful life in the Air Force, many of the Super Constellations were transferred to Air National Guard (ANG) units in various states. Most started receiving their Constellations in 1962. Some ANG Connie operations are listed below:

District of Columbia The District of Columbia ANG unit is unique in that its home base is not a state. Further, it is a fighter squadron, not a transport organization. However, it operated one C-121C as a support plane from May 1957 into 1969.

Mississippi The Mississippi ANG unit received its first C-121C in July 1962 and eventually operated a total of eight until April 1967. These were used for transportation, evacuation, and support duties over the North Atlantic and in the United States.

New Jersey The New Jersey ANG flew nine C-121Cs and one C-121G on transport and aeromedical duties from McGuire AFB from 1962 until 1973. This unit was unique in that it was the last ANG unit to fly Super Constellations in the medical transport role.

Pennsylvania Pennsylvania's ANG was unique in having two separate divisions. One was stationed at Middle-Olmstead Field in Harrisburg while the other was stationed at Pittsburgh. The unit in Harrisburg used fourteen C-121Cs and one C-121G at first for medical transport duties but later in support of MATS/MAC (Military Air Transport Service/Military Airlift Command) transport operations. The unit was redesignated in 1958 as a Tactical Electronic Warfare Squadron and thus acquired four EC-121s. Retirement came in early 1967 for the C-121Cs and the C-121G, but the EC-121S models remained in service until 1979 and were the last Super Constellations to be operated by any ANG unit.

Meanwhile the unit in Pittsburgh operated twelve C-121Gs from 1962 until 1972 in passenger, medical evacuation, and cargo roles to all parts of the world. This unit had the unique distinction among ANG Connie operators of being activated to full MAC status during the Vietnam War. A unit redesignation in 1972 meant that its Connies would be retired and by late 1972 all were in storage.

West Virginia The West Virginia ANG also operated two models of the Super Constellation, the C-121Gs from 1963 until replaced by the C-121Cs in 1967. They lasted until 1972 and were used for carrying passengers and cargo throughout the Caribbean area and the United States.

Wyoming The Wyoming ANG had the distinction of operating only the C-121G. It operated twelve, also for medical evacuation, cargo, and passenger transport duties, from 1963 until 1972, when they were replaced by the C-130B.

ANG Headquarters The headquarters of the Air National Guard based three C-121As and one C-121C at Andrews AFB, Maryland. These planes were used for staff transport and administrative duties from early 1967 until being retired in mid-1975.

Air Force Reserve Super Constellations

Other units to receive Constellations were those of the AFRes, including the headquarters section. This section unofficially operated one C-121C from mid-1971 until late 1972. The only other AFRes unit to receive Constellations was the 79th AEW & CS (Airborne Early Warning and Countermeasures

Squadron) at Homestead AFB, Florida. Overall, they operated a total of twenty-six Constellations made up of two C-121Gs, nine EC-121Ds, and fifteen EC-121Ts. The unit was equipped with the EC-121Ds in 1971, but these were gradually replaced by the EC-121Ts as they became available. These airplanes, along with the C-121Gs, remained in service until 1978 when they were gradually retired.

Other Military Super Constellations

Although the USAF and Navy were the only original purchasers of military versions of the Connie, two other nations, India and Indonesia, acquired former airliners for military or naval service.

India

The Indian Air Force acquired two 1049Cs, three 1049Es, and four 1049Gs from Air India in 1961–1962. Two were already equipped with cargo doors so were used as long-haul transports while the rest were converted for search and rescue and maritime reconnaissance missions. For these they were fitted with new avionics along with other special equipment. They replaced the Consolidated B-24 Liberator bombers that had previously been used for this purpose. In 1976 the

search and rescue mission was transferred to the Indian navy, along with five of the air force Super Connies. These were retired in 1984, a date that made them the last Constellations in the world in military service.

The Indian military Super Connies are listed here in sequence of their air force serial numbers matched to Lockheed model and serial numbers, Indian civil registrations, and Indian navy serial numbers.

Indonesia

Indonesia, which gained its independence from Holland after World War II, acquired the last three Pakistan International Airlines Super Connies, one 1049G-55 and two 1049Hs. Pakistan donated the planes to the Indonesian air force as a token of gratitude for Indonesia's assistance during Pakistan's war with India.

The match of the Indonesian air force serial numbers, T1041–1043, to former civilian registration numbers is not known. Lockheed and Pakistani data for the three airplanes are as follows:

Lockheed Model No.	Lockheed Constructor's No.	Pakistan Civil No.
1049C-55	4522	AP-AFS
1049H	4835	AP-AJY
1049H	4836	AP-AJZ

Indian AF Serial No.	Lockheed Model No.	Constructor's No.	Indian Civil No.	Indian Navy Serial No.
BG575	1049E-55	4614	VT-DHM	IN315
BG576	1049G-82	4666	VT-DIM	IN316
BG577	1049E-55	4615	VT-DHN	—
BG578	1049G-82	4646	VT-DIL	—
BG579	1049G	4687	VT-DJX	—
BG580	1049E-55	4513	VT-DHL	IN317
BG581	1049C-55	4517	VT-DAL	IN318
BG582	1049C-55	4548	VT-DGM	IN319
BG583	1049G-82	4686	VT-DJW	—

The Ultimate Constellation: Model 1649A Starliner

As with the original Model 049 Constellation and the 1049 Super Constellation, TWA was the prime mover in bringing about the 1649A. The airline asked Lockheed to produce an even longer range version of the 1049 to meet the challenge of the new Douglas DC-7C ordered by the competition. This had been named Seven Seas for its transoceanic range, thanks to a 10ft increase in wingspan and a resulting increase in aspect ratio that lowered induced drag. The DC-7C wing also had room for more fuel. The aircraft was capable of nonstop transatlantic flights in both directions year-round,

and TWA needed a new airplane with which to compete.

In May 1955 work began on the 1649A, which at first continued the Super Constellation name, then became the Super Star Constellation, and finally in March 1957, the Starliner. The result of this program was that the piston-engine airliner was brought to the peak of its development.

Lockheed responded to the challenge of the DC-7C with the new Model 1649A (there was no plain Model 1649, nor was there to be any subsequent development). Basically, the 1649A put an entirely new wing on the Model

1049G fuselage, tail, and power package. The new wing was built as a single unit and was installed in a cutout under the fuselage instead of attaching to center-section wing stubs as previously.

The new wing had a span of 150ft compared to the previous 123, and 1,850sq-ft of area. The planform was changed to a leading edge with very slight sweepback and a highly efficient aspect ratio of 12:1. A major change was to thinner airfoils, the symmetrical NACA 0015 at the root and the NACA 0011 at the tip. In the face of latter-day laminar airfoil development, this was a

First-takeoff photo of the Model 1649A on October 11, 1956, emphasizes the increased span, new planform, and wider engine spacing of the new model compared to the old. Lockheed

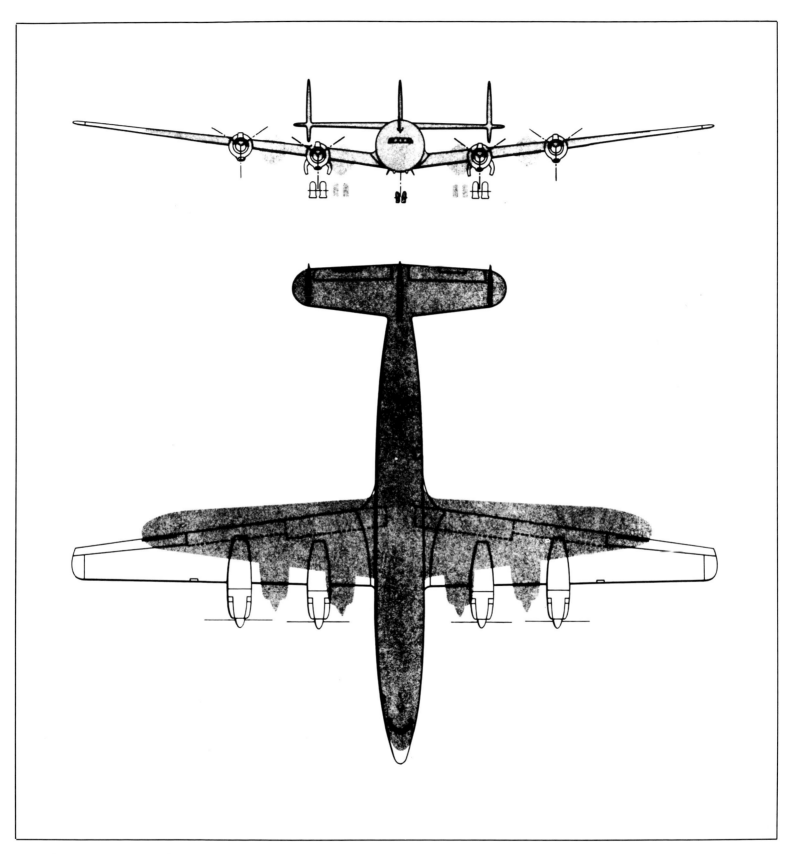

This drawing compares the sizes of the Model 1649A wing and that of the Model 1049. It also shows the additional area between the inboard engines and the fu- selage. The 1649A was also available with the weather radar that had been adopted on the 1049G. Lockheed

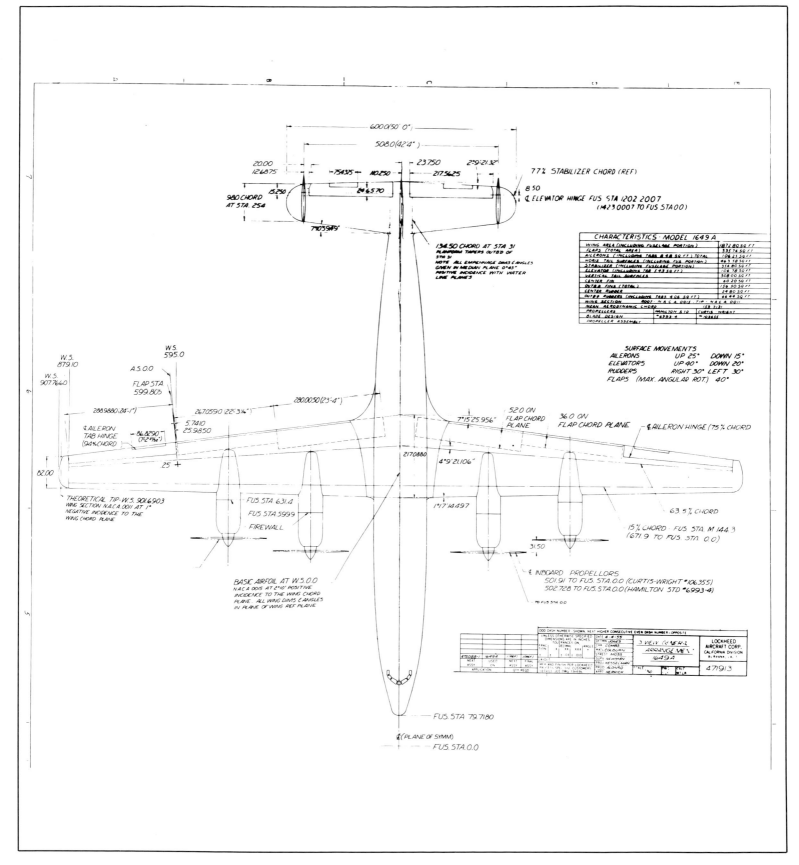

*Lockheed three-view drawing of the Model
1649A. Lockheed*

big step into the past. The Boeing B-17 of 1935, the Boeing 307 Stratoliner of 1938, and the Boeing 314 Clipper of the same vintage all used the symmetrical NACA 00-series airfoil. As on the 1049C, tip tanks were optional on the 1649.

The new wing was a boon to the passengers in that the noisy turbo-compound engines were now mounted 5ft farther outboard and propeller noise was further reduced by lower gearing of the 3,400hp 988TC18EA-2 engines. Cabin noise was also reduced by an additional 900lb of soundproofing material.

Anachronistic airfoils aside, the new wing put the 1649A ahead of the DC-7C. Fuel capacity was increased from the 1049G's 4,760gal to 9,278gal.

This increased the range and permitted nonstop polar flights from Los Angeles to London, which usually took nineteen hours eastbound. One westbound flight from London to San Francisco took twenty-three hours, nineteen minutes.

While Lockheed named the new model Starliner, TWA advertised its twenty-nine examples as Jetstream, either to indicate a very high cruising altitude or to suggest that its customers were going to board a jet.

Two different types of propeller were offered with the Starliner: the normal, solid-blade propeller, and a hollow-blade propeller (saving 680lb per aircraft) costing nearly twice as much. TWA initially chose the hollow blades but was forced to replace them

when problems arose with the nylon foam core material.

The first 1649A, appropriately registered N1649, flew on October 10, 1956, and as 1649A-98 the plane got a new ATC, 4A-17, on March 19, 1947. (The 1649A was built in the same plant as earlier models. The change from 6A to 4A in the ATC number reflected a renumbering of the FAA regions.) It also backed up to get a new c/n series, 1001–1045. Gross weight was now 160,000lb and the allowable landing weight was raised to 123,000lb. Speed was 377mph at 18,600ft, cruising speed was 290mph, and the landing gear could be used as a speed brake up to 269mph. Range was 4,940 miles with a 19,500lb payload and 6,180 miles with an 8,000lb load. The unit price was $2.9 million.

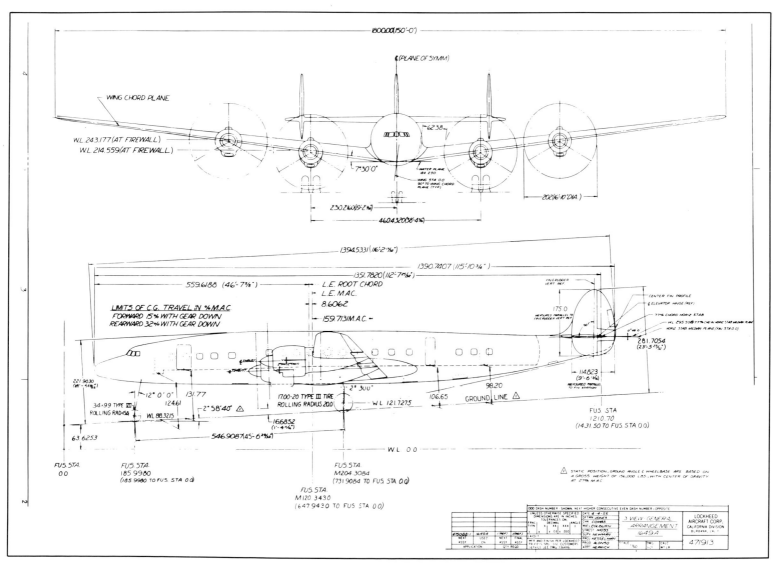

Lockheed claimed that the 1649A had the greatest range of any airliner of the time and was able to fly from New York to Paris in nearly three hours' less time than the DC-7C when carrying the same payload. It also claimed that the Starliner was 70mph faster than any other piston-engine airliner at ranges over 4,200 miles and was capable of bringing every European capital within nonstop reach of New York. This was indeed true, since the Starliner could carry a payload of 17,000lb over 5,300 miles while cruising at 350mph. As with any airliner, however, range and payload were trade-offs; to increase one it was necessary to decrease the other.

TWA was the major Model 1649A operator, having ordered twenty-five and later acquiring four others, and placed its first Jetstreams in service in late May 1957. Air France took delivery of ten and, in an impressive display of its range, flew its first Starliner 5,800 miles nonstop over the North Pole from Los Angeles to Paris in a record-breaking seventeen hours, eleven minutes. The German airlines Lufthansa, which ordered four, was to better this record

The wing of the Model 1649A was built and installed as a single unit instead of having inboard sections attached to center-sec- tion stubs. Lockheed via San Diego Aerospace Museum

Foreshortening by the long lens used for this flight photo of the first Model 1649A deemphasizes the increased wingspan, but shows off the new wing planform and the increased distance between the inboard engines and the fuselage. Lockheed

when its first 1649A arrived in Hamburg. The airplane took off from Burbank and completed the 7,000 mile trip nonstop in seventeen hours, nineteen minutes.

Another European airline that had placed an order for four Starliners was Linee Aeree Italiane (LAI) of Italy. LAI merged with Alitalia, which had ordered DC-7Cs, in November 1957, so the Starliners it ordered were added to TWA's fleet. An airline that had also intended to operate 1649As was Brazil's Varig airline, which ordered two in June 1956 for its Buenos Aires—New York route. This order was later changed to a repeat order for three 1049Gs, as Varig wanted to standardize on this version.

The US Navy was interested in producing an AEW version of the Starliner to be designated W2V-1. They were to be equipped with rotodomes as on the WV-2E and were to have Allison T-56 turboprops replacing the original Wright turbo-compound piston engines. Because the projected maximum takeoff weight of 175,000lb was more than the Allisons could handle alone, two Westinghouse J-34 turbojets were to be mounted in wing-tip pods. Lockheed had completed design studies on such an aircraft, but military budget restraints forced its cancellation.

After the Starliner was introduced, the user airlines were able to establish new routes because of its superior range. On October 2, 1957, TWA inaugurated the second over-the-pole service between Los Angeles and London, with a stop at San Francisco. The first had been started by Pan Am three weeks earlier with DC-7Cs, but TWA's service was considerably faster. It is on this route that the Starliner still holds the record for the longest continuous piston-engine airline flight of twenty-

Although basically unchanged from the Model 1049, the cockpit of the 1649A has additional instruments and controls associated with the upgraded airplane. The fuel dump valve controls are visible on the ceiling. Lockheed

With emphasis on range rather than payload, the Starliner normally carried fifty-eight to sixty-four passengers in four-abreast first-class seating on international *routes. Up to ninety-nine could be carried on domestic routes, with a typical compartmented interior seating twenty-six first-class and forty-five tourist class. Lockheed*

three hours, twenty minutes, set in October 1957. Air France also inaugurated its own over-the-pole service starting on April 10, 1958. The route linked Paris with Tokyo and required a stop at Anchorage, Alaska, for refueling.

The Starliners had been in airline use for only a couple of years when they began to be replaced by the new jets arriving on the scene. Because of this, and also because the airplanes were so new and still had to pay for themselves, some airlines had their Starliners converted into freighters. TWA had twelve converted and Lufthansa had two of its original four converted. These conversions ensured that the airplanes would be in use much longer, and as such, they were among the last piston-engine airplanes used by these airlines.

With the jet age coming on strong, only forty-four 1649As were built (c/n 1043 was canceled.) While it was unquestionably the best of the Connies, the 1649A had, thanks to the new jetliners, the shortest trunk-line career of any Connie model.

The last of the Lockheed Constellations was delivered on February 12,

As a substitute for sleeping berths, Model 1649A Starliners (and some earlier models) could be fitted with special *"sleeper seats" that permitted the passengers to stretch out farther than in ordinary day seats. TWA via Ed Peck*

One compartment of the Model 1649A Starliner could be fitted out as a flying cocktail lounge. (Note: All of the photos in this book showing passengers in a Lockheed Constellation feature professional models who were paid by the company. In the 1930s such setups were less costly; a few suitably dressed employees were rounded up from the offices and put aboard the airplane to be photographed.) TWA via Ed Peck

1958, to end a sixteen-year production life, the longest for any piston-engine transport. By 1960 the major airlines were assigning their nearly new 1649As to their secondary routes or converting them to freighters.

The Model 1649A, best and last of the Constellations, was a thoroughly good airliner but it appeared too late, a full year behind the DC-7C, to become a major player on the world airline scene.

TWA was the major user of the Model 1649, having bought twenty-five new and later acquiring four used models. Company-originated name "Jetstream" appears on the nose. E. M. Sommerich

Lufthansa's Model 1649A Super Star Constellations were used on routes from Germany to the United States, Canada, and South America starting early in 1958. Normal passenger seating was sixty-two, but there was a special thirty-two-passenger ultra first-class interior for the special New York City "Senator" service. Lockheed

From the start, TWA's "Jetstream Starliners" took over most of the longer transcontinental and North Atlantic services from the older Super Constellations. Even though TWA's Starliners were equipped with a basic interior for sixty-eight passengers and a crew of eleven, over fifty different interior variations were used during its career. ATP/Airliners America, Arthur Muhler

The Model 1649A featured an entirely new wing on the basic Model 1049 fuselage. The wing was longer with squared wing tips and the engine nacelles were re-designed. This is the second 1649A-98 built, and was delivered to TWA as N7301C, its first Jetstream, on September 4, 1957. Lockheed

Air France 1649A F-BHBK on a factory test flight before delivery on June 30, 1957. Passenger comfort in the 1649A was en-hanced by the quieter cabin that resulted from moving the engines five feet farther outboard than on previous models. Lockheed

Independent North American Constellation Operators

In the 1950s, 1960s, and 1970s many second-tier Constellation operators existed throughout the world that were not scheduled airlines. These included charter operators, private businesses, and other aerial service providers. The highest concentration of independents was in the United States. Therefore this chapter covers most of the US operators, with Canadian operators included.

United States

The second-tier independent operators are listed alphabetically by groups according to the part of the country in which they were based. Not all were exclusively Connie operators. While some were poorly financed one-airplane operators, others were well established, with some using Connies in a larger mixed fleet or eventually replacing them with later equipment.

One of two former LAV (Venezuelan) Model 049-46 Constellations acquired by Braniff, a major US airline, in August 1955. They were sold off in 1960 and 1961. E. M. Sommerich

One of two Model 1649A-46s leased by Alaska Airlines for its MATS flights from the United States to Europe. Alaska also had three leased 1049Hs for service within Alaska and to the United States. Norman E. Taylor

The single Model 049 used by California Hawaiian Airlines (CHA) in 1952. This was originally C-695 42-94559 that went *to BOAC as G-AHEN. It went to CHA as N74192 in 1952. William T. Larkins*

Airlift International took over the rights and assets of the defunct Slick Airways in June 1966 and thus acquired six 1049Hs. Another 1049H was leased in November and all were used for Military Airlift Com- *mand (MAC) charter work from the company's Miami base until being retired at the end of 1967. ATP/Airliners America, John Stewart*

Several major US scheduled airlines, such as Braniff Airways and Capital Airlines, that were not first-tier Connie operators acquired used airplanes from other firms and added them to their existing fleets. Such operators are not covered in this chapter.

It should be noted that some operators held their Connies for short periods of time—sometimes only a few months—and that others owned Connies but never operated them.

Alaska

Alaska was home for five different Connie operators, most of them established to cash in on the famous (or infamous) North Slope oil boom. A sixth Connie operator, Alaska Airlines, was a long-established scheduled airline with a route to Washington and Oregon, and also held MATS contracts for flights to Europe.

Interior Airways operated out of Fairbanks as an air-taxi with a leased 1049H from January through July 1966. With the hope of increased business from the North Slope oil exploration, it leased three more 1049Hs in 1969 but made little use of them before they were returned.

North Slope Supply Co. was another operator dependent on the oil boom. Operating out of Anchorage, North Slope purchased nine 1049Hs in April 1969 to operate supply flights, but only seven were ever delivered before numerous delays forced the company into bankruptcy.

Prudhoe Bay Oil Distributing Co. purchased three 1649s in December 1968 for use as bulk fuel oil carriers to the North Slope. Flights were carried out from Anchorage before a temporary halt on construction of the trans-Alaskan pipeline resulted in the company's bankruptcy by late 1969.

Wein-Air, Inc. was a small charter company based at Bethel that purchased two 1649s in June 1973, and used one for freight flights within Alaska. By October the airplanes were out of use and were sold, but the company purchased three other 1649s in September 1974, with which to resume operations. Two were made airworthy, but official approval was not forthcoming and all were sold in November 1975.

California

The state of California was even with Florida in the number of independent Connie operators based there, with twenty-one. Some were well-established organizations that operated their Connies prosperously for many

Bal Trade operated two 1049Hs on cargo flights from Florida to various destinations in Central and South America from late 1968 until the end of 1970. ATP/Airliners America, John Stewart

Central American Airways Flying Service began using a 749A from Kentucky in October 1967 on general charter work until it was sold in March 1973. A 1049H had joined the fleet in January 1969 and was used for freight services for the automobile industry until October 1978. ATP/Airliners America, Bruce Drum

A former Eastern Air Lines Model 649-79-12 converted to a 749, N105A, was leased to Trans California Airlines in 1964.

The forty-five-degree black and white tail decoration was quite unique. J. Roger Bentley

One of the two Model 049-46s operated by Edde Airlines of California. Both were leased from TWA in 1962 and 1963. The window pattern was revised from original C-69 and Model 049 airliners. Peter M. Bowers collection

A former QANTAS Model 1049E-55-01 was returned to Lockheed ownership in 1959 and was then leased to Twentieth Century Airlines before being leased to Trans International Airlines as N9719C. Walter J. Redmond

Capitol Airways of Nashville operated a total of eighteen Constellations at one time and was the largest nonscheduled operator of the type. Overall, Capitol operated twenty-three different Constellations, including four 749As, five 1049Es, two 1049Gs, and twelve 1049Hs. Its first Constellation, a 749A, arrived in early 1957, and all were used on civilian and MATS charters until being put into storage in 1968. ATP/Airliners America, Bernard Crocoll

years while others were short-lived marginal operations.

Admiral Airways began operating Constellations in June 1961 when two 749As were leased. A 1049 was added that year along with three more 749As in April 1962 for use on passenger charters, but the company ceased operating in October.

California Hawaiian Airlines started low-fare service from Los Angeles and San Francisco to Honolulu with a single 049 in 1952. These flights continued until October 1953, when the carrier suspended operations. It reemerged in 1955 but didn't begin using Constellations again until December 1960, when two 1049s were leased. Another 1049 was added in 1961 and all three were used on regular flights to Europe from June until October. Also that year, four 749As were added to the fleet to provide equipment for domestic charters before the carrier ceased operating in 1962.

Dellair of Long Beach used three 049s for flights within California in 1964 before the company dissolved.

Edde Airlines of North Hollywood began operations in September 1962, with a leased 049. Another 049 was added in 1963. Both were used for domestic charters and military contract work before the company ceased operations in 1966.

Futura Airlines, an Oakland-based carrier, used two 049s during 1962 on services to Lake Tahoe but halted all operations within the year.

Great Lakes Airlines of Burbank purchased eight 749As in 1961, although many were subsequently leased to other operators. They did operate a few themselves within California before the airline folded in January 1962.

Hawthorne Nevada Airlines began operating from the Los Angeles area to Hawthorne, Nevada, in June 1968, with one 049. The following April its name was changed to Air Nevada, and services to Lake Tahoe, San Jose, and Oakland were offered before operations ceased in September 1969.

The Hughes Tool Co., or Howard Hughes depending on how you want to look at it, owned many different types of Constellations throughout its operation. These included four 049s from 1946 until 1950, a 749A from 1951 until 1954, a 1049G from 1956 until 1960, and a 1649A Starliner in 1957.

Lake Havasu City Airlines, although based in Long Beach, used five 049s starting in February 1964, to ferry potential buyers to the newly built city in Arizona. These flights continued until early 1970, when the Constellations were replaced by later Lockheed Model 188 Electras.

Mercury General American Corporation of Torrance used five 1049Hs on general charter work and also under contract to MAC starting in March 1966. These services were short-lived, however, as the company stopped operating in 1967.

Midair of Barstow purchased one 1049G and six 1049Hs in April 1973 for conversion to sprayers, but declared bankruptcy before any conversions were started.

Pacific Air Transport operated out of Santa Rosa with one 049 and one 749A from 1968 until both were put into storage in 1969 and 1970, respectively.

Paramount Airlines, based in Burbank, used three 749As and one 1049 for low-fare transcontinental flights from March 1961 until the aircraft were withdrawn in early 1962.

South Pacific Airlines of San Francisco used two 1049s from Honolulu to Pago Pago, American Samoa, starting on March 14, 1962. This route

Braniff International Airways purchased two 049s in August 1955 from LAV for use on services within Texas and also to Chicago and Kansas City. Braniff also operated an interchange with TWA between Houston and the West Coast via Dallas and Amarillo until early 1958 when the airplanes were put into storage. J. Roger Bentley

A Houston-based airline, CJS Aircargo, purchased a Model 749A and a Model 1649A in December 1970. These were used on freight flights within the United States and also to Central and South America before being put up for sale in 1972. ATP/Airliners America, Bruce Drum

This Model 1049H, N1927H, was sold by Lockheed to Air Finance Corp. on May 17, 1957, and that firm leased it the same day to Transocean Airlines. It eventually ended up as CF-AEN in Canada. The airplane had notably fewer windows than other passenger Super Constellations. Lockheed

Trans-Canada Airlines bought this Model 1049E-55 in June 1954 as CF-TGF. After use by several US operators as N9742Z it was acquired by American Flyers in August 1964. David W. Lucabaugh via Norman E. Taylor

Hawthorne Nevada Airlines began operating from the Los Angeles area to Hawthorne, Nevada, in June 1968 with one 049. In April 1969 the name was changed to Air Nevada, and services to Lake Tahoe, San Jose, and Oakland were offered before the airline ceased operating in September 1969. ATP/Airliners America, Jack Bol

Alaska Airlines lease-purchased two 1649As in early 1962 for its MATS contract flights to Europe. They also purchased another 1649 that was used for spares only. Alaska later leased three 1049Hs for scheduled services within Alaska and also to Seattle and Portland. The 1049Hs were returned in 1966 and 1967 and the last 1649A was withdrawn in September 1968. ATP/Airliners America, Preston Foreman

was the longest overwater route flown by a 1049 before it was abandoned in early 1964.

Standard Airways in San Diego, a nonscheduled carrier, operated specialized flights from San Francisco and Burbank to Honolulu in 1962. A large fleet of Connies was acquired for this route, including three 049s, five 749As, one 1049, three 1049Gs, and one 1049H. The loss of a 1049G in May 1963 significantly affected the airline and it voluntarily ceased operations in February of the following year.

Trans California Airlines of Burbank offered low-fare intrastate services with a fleet of six 749As starting in July 1963. These airplanes were fitted with ninety-eight seats, the most ever put into a short-fuselage Connie. The crowded seating may have contributed to Trans California's demise. It ceased operations by the end of 1964.

Trans International Airlines operated a large fleet of Connies on freight charters and military contracts from its Los Angeles base. Its first Connie was acquired in December 1960 and the fleet grew to include one 749A, three 1049Gs, nine 1049Hs, and at least one 1049E until they were put into storage in mid-1967.

Transocean Airlines of Oakland purchased one 1049H in July 1957 for low-fare services from Oakland and Burbank to Honolulu. Another five 749As were purchased in January

1958, along with another 1049H in April 1959 in support of military contracts until being replaced by transports later that summer. The company declared bankruptcy on July 11, 1960.

Twentieth Century Airlines leased six 1049Gs briefly in late 1959 and early 1960 for use on MATS transpacific charter flights.

World Airways leased five 1049Hs in early 1961 for its MATS contract flights from California to Bangkok, Tokyo, Okinawa, and Manila. Two more 1049Hs plus four 1649As were added to the fleet in mid-1962. They were used on domestic interstate passenger and freight charters before the airplanes were returned to their lessors in 1963 and 1964.

World Wide Airlines operated out of Burbank with two 1049s starting in July 1960, and added two 749As in early 1961. These planes were used on military contract flights along with civilian charters until the military stopped using World Wide in November 1961. All services were suspended until July 1962, when a fleet of four 049s was purchased, but this was short-lived and the carrier was out of operation by October.

Central States: Illinois, Kentucky, Indiana, and Tennessee

Only one Connie operation was based in each of these four states. Note that while one was based in Illinois, it operated out of Miami.

Capitol Airways of Nashville, Tennessee, operated a total of eighteen Constellations at one time and was the largest nonscheduled operator of the type. Overall, Capitol operated twenty-three different Constellations, including four 049As, five 1049Es, two 1049Gs, and twelve 1049Hs. Its first Constellation, a 749A, arrived in early 1957, and all were used on civilian and MATS charters until being put into storage in 1968.

Note: Capitol Airways should not be confused with the scheduled airline Capital Airlines of Pennsylvania, another second-tier Connie operator.

Central American Airways Flying Service began using one 749A from Kentucky in October 1967 on general charter work until it was sold in March 1973. One 1049H joined the fleet in January 1969 and was used for freight services for the automobile industry until October 1978.

Passaat Airlines of Schiller Park, Illinois, operated one 1049 and one 1649 starting in early 1965 on charter flights out of Miami. These flights continued until the Starliner was destroyed in an accident in December 1966. The company ceased operations by early 1967.

Raitron, Inc. operated one 1049A from Gary, Indiana, starting in June 1970 and continuing until the airplane crashed in September 1973.

East Coast: North Carolina and Washington, DC

Air America, based in Washington, DC, owned one 1049H. The aircraft was based in Taiwan and operated throughout Southeast Asia from July 1963 until March 1969.

Blue Bell, the company that owns Wrangler Jeans, operated two 1049Hs out of Greensboro, North Carolina, starting in early 1969. They were used to fly raw materials and finished garments between North Carolina and Florida or Puerto Rico until the airplanes were sold in 1973.

Florida

Florida was home to a total of twenty-three independent Constellation operators. Most were based in the Miami area with another large group at Fort Lauderdale. Because of the nature of some of the Miami operations, the portion of the airport assigned to them earned the nickname "Cockroach Corner."

Aeroborne Enterprises of Fort Lauderdale purchased three 1649As for cargo charter services in 1980, but nothing became of the venture.

Aerolessors of Miami acquired one 1049H in July 1966 for freight charters and added a second 1049H the following March. A third 1049H was leased in November 1968, before all services were stopped in 1970.

Aero Sacasa of Fort Lauderdale purchased one 1049 in November 1976 and one 049 in April 1979, but neither airplane was flown by the company.

Air Cargo Support of Miami used four Connies, one 1049G and three 1049Hs, on charters carrying cattle to Central and South America starting in September 1973. By 1976 only one of the 1049Hs remained flyable. It, too, was retired in 1982.

Air Fleets of Fort Lauderdale purchased one 1049 and one 1049G for charter work in January 1970. They were used until the company ceased operations in early 1971.

Airlift International of Miami took over the rights and assets of the defunct Slick Airways in June 1966, and thus acquired six 1049Hs. Another 1049H was leased in November, and all were used for MAC charter work from the company's Miami base until being retired at the end of 1967.

ASA International, based in St. Petersburg before moving to Miami, leased three 049s for charter work within California in 1962 before they were returned the following year. An attempt was made to restart operations with the 1649As reengined with turboprops in 1965, but nothing became of the project.

A former US Army C-69, 43-10312, delivered to Capital Airlines in September 1950 as N67593 and leased by them to Imperial Airways in 1960. ATP/Airliners America

Associated Air Transport of Miami, a nonscheduled and supplemental carrier, acquired four 749As in late 1961 and early 1962 for possible use on MATS contracts. However, the military disqualified the carrier and its operating certificate was revoked in October 1962.

Aviation Corporation of America, based in Fort Lauderdale, purchased eleven 1049s in late 1968, followed by two 1049Cs in 1969, for planned intrastate services. All the airplanes were either sold or scrapped before services began.

Bal Trade operated two 1049Hs on cargo flights from Florida to various destinations in Central and South America from late 1968 until the end of 1970.

Carib Airways of Fort Lauderdale purchased one 749A in January 1979, but the airplane was subsequently seized for suspected drug smuggling.

F & B Livestock operated one 1049H out of Hialeah from early 1971 on cargo charters carrying meat to Central and South America until the airplane was damaged in May 1976.

Florida State Tours of Miami purchased one 1049 in June 1963, and another five in August 1964 for intrastate services. Ultimately, only one of the five was ever delivered and commercial services were never started before the airplanes were sold in 1964.

International Caribbean Corporation purchased one 049 in August 1965 for charter flights to South America, but it was impounded a month later for smuggling.

Lloyd Airlines briefly operated two 049s out of Miami in mid-1961 until one was impounded for smuggling and the company ceased operations.

Magic City Airways was a small contract-carrier from Miami that used two 049s from January 1962 until the company stopped operating in 1965.

Miami Airlines leased four 749As for passenger charters in mid-1960 but stopped operating by late 1961.

Modern Air Transport operated five 049s from Miami on intrastate charters and cargo flights starting in August 1962. Two 1049s were added, one in 1961 and another in 1963, along with a 749A in 1964, before all were retired in 1965.

Sky Truck International of Fort Lauderdale operated two 1049Hs from September 1972 on cargo charters to Central and South America before all services were stopped in 1975.

Transair Cargo of Miami used one 1049H for cargo charters from 1972 until the company ceased operating in November 1977.

Trans American Leasing used three 1649As from May 1968 on a contract cattle flight to Algeria and also on flights to South America. All three Starliners were out of service by 1970, so one 1049 was purchased and continued in use until 1973.

Vortex used three 1049Hs for nonscheduled charter work from October 1971 until ceasing operations in April 1973.

Hawaii

The island state of Hawaii had only one Connie operator.

Willair International of Honolulu purchased one 1649A in August 1968

The former Transocean Model 1049H in its final configuration, registered to Hellenic Air Ltd. of Canada as CF-AEN. The cabin windows were reduced in number even more from the transocean arrangement. John Wegg

The Hacienda Hotel of Las Vegas, Nevada, operated a fleet of six secondhand Constellations for charter flights and to transport its guests. This is former TWA 049-46-25 N86517. Victor D. Seely

to transport leather from Barcelona, Spain, to Taiwan, and then the finished products to the United States. The original 1649A was involved in a training accident within a month of its purchase, so another 1649A was acquired in December and used until the company folded in 1970.

Northeastern States: Connecticut, New Jersey, and New York

These three states hosted nine of the independent Connie operators.

Air Mid East of Syracuse, New York, operated one 1049H in July 1968 on a trip to Madrid, Spain, and also attempted to fly supplies to Biafra, Nigeria, just before the company dissolved.

Burns Aviation of Stratford, Connecticut, used two 1649s in 1976 for livestock charters to South America and Asia.

Coastal Air Lines of West Trenton, New Jersey, operated four 1049s between 1958 and 1962 as a nonscheduled and supplemental carrier.

Colonial Airlines of New York City leased several 749As in the summer of 1954 for use on its daily New York–Bermuda route, but by June 1956 it was taken over by Eastern Air Lines.

Flying W Airways of Medford, New Jersey, purchased two 1649As in November 1968 but sold them within the month. It purchased two 1049Hs shortly afterward, but they didn't last very long either, since the company went out of business in mid-1971.

Intercontinental U.S. was another New York City-based operator that started passenger and freight charters in May 1962 with two 1049Hs. Two more 1049Hs were added for the 1963 summer season, but the company ceased operating on March 17, 1964.

Moral Rearmament Corp., a religious-political group based in New York City, used one 1649A from June 1965 until early 1966 as a corporate airplane.

Starflite of White Plains, New York, used one 1649A from 1965 until 1966 on specialized US domestic and international contract services for major corporations.

U.S. Airlines of New York City leased one 049 during 1951 and 1952 for nonscheduled services from Miami to New York.

Northwestern States: Oregon and Wyoming

Oregon and Wyoming were home to only one Connie operator each.

Conifair Aviation acquired two C-121As in January 1980 to conduct spraying on behalf of the Canadian Department of Lands and Forests. This continued each year until the airplanes were put up for sale in 1985. ATP/Airliners America, Russell Brown

Christler Flying Service of Thermopolis, Wyoming, used five former USAF C-121As for sprayers from July 1970 until they were sold in April 1979. During this period they flew many spray missions all across the United States and Canada.

General Airways in Oregon leased a 1049 in November 1960 for passenger and freight charters on transatlantic flights. They didn't last very long, however; the carrier was out of business by the following March.

Southwestern States: Arizona, Nevada, and Texas

These Southwestern states hosted eight independent Constellation operators, one of which had one corporate name but operated under another.

Aircraft Specialties of Mesa, Arizona, was that state's first Connie operator. It traded under the name Globe Air and offered crop spraying with two 749As and a former military 1049 starting in September 1972. In early 1973 one 1049G and three 1049Hs were added to the fleet, plus an additional 1049H in August 1974. The airplanes were used within the United States and Canada. Some were put in storage by summer 1975, but others continued in use until 1981.

American Flyers Airlines, based in Fort Worth, Texas, purchased four 049s in April 1960 for domestic passenger and freight charters plus military contracts. The airline added overseas charters and therefore acquired four 1049Gs in mid-1964, plus an additional 1049G. By 1967 the 049s had been withdrawn, and the 1049Gs were stored within a year.

Belezean Airlines, based in Houston, Texas, used one 049 for charters from March 1967 until June 1969.

CJS Aircargo of Houston, Texas, purchased one 749A and one 1649A in December 1970. The aircraft were used on freight flights within the United States and also to Central and South America before being put up for sale two years later.

Globe Air was the operating name for Aircraft Specialties of Mesa, Arizona.

Las Vegas Hacienda Hotel was Nevada's only Connie operator, using five 049s and one 749 on charter flights starting in August 1959. The services were mainly from the West Coast and continued until March 1962, when the airline stopped operating.

Produce Custom Air Freight of Phoenix, Arizona, purchased one 049 in July 1972 and had it delivered in March 1973, but it remained unused until it was sold in 1976.

S. S. & T. Aerial Contracting of Buckeye, Arizona, purchased one 049 in December 1972 for conversion to a sprayer. The airplane remained unmodified until it was sold in 1977.

Unlimited Leasing was an airline that started in Houston, Texas, but moved to Hialeah, Florida. It purchased one 749A in November 1970 and used it quite extensively before selling it in January 1979.

Federal Aviation Administration (FAA)

In 1964 this government agency bought a former TWA 1649 Starliner, serial N7307C, c/n 1008, for the specific purpose of crashing it. The object was to provide data on passenger and crew seat strengths and restraint systems. Cameras were installed inside the airplane along with twenty-one dummy passengers, while external cameras recorded the impact.

On September 3, 1964, the Starliner was guided to the impact area by rails and, to its credit, was airborne for just a second before impact. Afterward, the fuselage was used for mock emergency evacuation drills with local people playing the parts of the injured passengers.

The FAA also acquired both former Navy WV-1s, but soon turned them over to the USAF.

Canada

Canada made good use of second-tier Constellations, with a total of twenty-one being operated by six different organizations.

Beaver Air Spray purchased three 749s in April 1979 and used them for spraying in Quebec province before selling them to Conifair Aviation.

Canrelief Air was a Canadian relief organization that flew medicine and relief supplies to Africa during the Biafran War in 1962. Canrelief was the fourth owner of a 1049H registered CF-AEN, formerly N1927H. After its use on mercy flights, CF-AEN was acquired by another Canadian organization, Hellenic Air, Ltd., but was not put into service by that organization.

Conifair Aviation acquired two C-121As in January 1980 for use as sprayers on behalf of the Canadian Department of Lands and Forests. The C-121As were sold in 1985. Conifair also had three 749s converted to sprayers.

Downair of Newfoundland used one 1049H on cargo flights carrying fresh seafood to the United States. The short-lived service started in August 1973 and continued until the carrier's last flight on September 30, 1973, when the aircraft was stored.

Eastern Provincial Airways leased a 1049H during the latter part of 1968 to replace a lost airplane.

Montreal Air Services purchased a 1049G in 1964 and a 1049H in the following June. These aircraft were then leased to World Wide Airlines, which used them along with two more 1049Gs on freight and passenger charters. The flights continued until the carrier's operating certificate was revoked by the Canadian government in August 1965.

Constellations in Europe, Africa, the Middle East, and Asia

Europe

Europe was by far the major non-American user of second-tier Connies, with eight countries and eighteen operators involved as follows:

Austria

Aero-Transport Flugbetriebsgesellschaft purchased two 049s in June 1961 for passenger charters to holiday destinations in Europe. By May 1963, two 749As were purchased for freight charters but following allegations of gun-running, the airline was in trouble and ceased operations in the summer of 1964.

France

Air Inter used three 749As and one 1049G for regular domestic services from April 1961 until early 1962.

Catair was formed in late 1967 to provide passenger and cargo charters with four Model 1049Gs. Another two were acquired later, but all were gradually phased out of operation by early 1972.

Compagnie Air Fret, a non-scheduled operator, was formed in 1964 to provide passenger and cargo charters. Four 1049Gs were acquired in early 1968 but three were sold by late 1969, with the last being retired in August 1976, the last Connies on the European civil register.

Societe Aeronautique Francoise D'Affretement (SAFA) was a charter airline that used Constellations for a short time. It leased two 1049Gs from Air France in early 1966 for European charters until the end of 1967.

Iceland

Loftleidir used a 749A in 1960 for the increased summer traffic and a 1049H for a brief period in June 1963.

Ireland

Aerlinte Eireann Teoranta started service to the United States in April 1958. One 1049E, one 1049G, and two 1049Hs were leased and used on the Dublin–Shannon–New York route, with Boston added in October 1958. Aerlinte changed its name to Irish International Airlines in 1960 just prior to retiring the Constellations.

Luxembourg

Interocean Airways operated one 749A from June 1964. This airplane was used on passenger and freight charters until it crashed that October.

Although this Aerlinte Eireann 1049H-82 carries Irish Airlines markings, it has US registration N1009C because it is on lease from its first owner, Seaboard & Western. The airline name is in Gaelic on the left side of the airplane but is spelled out in English on the right side, a common practice. Frederick G. Freeman

Aerlinte Eireann succeeded in commencing service to the United States in April 1958. One 1049E, one 1049G, and two 1049Hs were leased and used on the Dublin–Shannon–New York route, with *Boston added in October 1958. Aerlinte changed its name to Irish International Airlines in 1960 just prior to retiring the Constellations. ATP/Airliners America*

This 1649A-98 was sold to Lufthansa in February 1958 as D-ALER. After leases to several other German and US airlines, it was leased to Luxembourg's Luxair in *April 1964, shown here still carrying its US registration N45520. It then became LX-LGX before going to South Africa as ZS-DVJ. ATP/Airlines America*

QANTAS was the original buyer of this 749-79-31 as VH-EAB in October 1947, and converted it to a 749A. After several resales and leases it was used by Ace Freighters *in England as G-ANUR. It went on the US register as N1949 and then to Argentina as CX-BHC before being scrapped. ATP/Airliners America*

Luxair reached an agreement with Trek Airways of South Africa in April 1964 to jointly operate 1649As on a weekly low-fare flight to Johannesburg. Luxair also operated a flight to London which connected with the Johannesburg flight at the appropriate times. The airplanes were retired in early 1969.

Nittler Air Transport International used one 1649 from August 1969 until May 1970 on charter flights.

Spain

Aviaco leased some 1049Gs from Iberia for passenger charters during the summer seasons of 1963, 1964, and 1965.

Inter-City Airways of Madrid used a 1049H in July 1968, but disappeared a short time later.

United Kingdom

Ace Freighters used eight 749s from March 1964 on all-cargo services throughout Europe and the Far East, but ceased operations by September 1966.

Euravia acquired three 049s in May 1962 and an additional four 749s. Two more 049s were purchased a year later, but all were replaced by turboprop Bristol Britannias by the summer of 1965.

Falcon Airways operated three 049s on charter flights in January 1961, but these operations led to trouble and the airline's operating certificate was revoked that September.

Lanzair first obtained a 749 in November 1973, but it was repossessed within a year. A 1049G was then purchased but it was impounded after being involved in arms smuggling. Another 749 had been acquired before this, but was unserviceable by November 1976.

Skyways of London purchased four 749As in July 1969 for all-cargo services, although passenger charter flights were introduced later. With the loss of an important contract in September 1969, all services were terminated and the Constellations were sold.

West Germany

Condor Flugdienst GmbH used two Starliners on specialized passenger charters from March 1960 until 1962. Two 1049Gs were leased from Lufthansa in 1964. They performed the

BOAC bought VH-EAB from QANTAS in February 1955, then leased it to Skyways of London, which converted it to a freighter as shown. Skyways sold it to Euravia in September 1962, and Euravia leased it to Ace Freighters in August 1965. ATP/Airliners America

same duties as the 1649s and remained in service until 1965.

Africa

Eight African nations and nine organizations operated Constellations. Only one country had more than one Connie operator, however.

Algeria

Air Algerie received two 749s from Air France for services to Paris starting in 1955, and also leased some 1049s before retiring them from service in 1961.

Burundi

Royal Air Burundi was formed in 1962 by a group of American business executives to provide nonscheduled

Air France bought 049-46 F-BAZA in June 1946. It was sold to TWA as 9412H in February 1950. TWA leased it to the African airline Royal Air Burundi as shown, still with its US registration, in December 1962. After many subsequent resales and leases, it ended up as a cocktail lounge in Greenwood Lake, New Jersey. Note the added weather radar nose and revised cabin window pattern. J. Roger Bentley

The former Luxair 1649A LX-LGX after sale to Trek Airways of South Africa as ZS-DVJ. The designation Super Star can be seen on the rear fuselage. The plane is now in a South African museum. ATP/Airliners America

services to Europe. The airline operated a few nonscheduled services with Connies and did some charter work before it halted all operations in 1963.

Ivory Coast

Air Afrique leased five Starliners from Air France in 1961 and later a 749 before the aircraft were withdrawn in 1962 and 1963.

Kenya

Britair East Africa of Kenya operated one 049 from 1964 on charter work before suspending operations the following year.

Former C-69-1 43-10313 was refurbished by Lockheed and sold to US operator Intercontinental Airways in 1947. After a period in Panama as RX-121, it returned to Lockheed for reconditioning and resale to the Israeli airline El Al as an 049 registered 4X-AKA in 1951. El Al converted it to a Model 149. Lockheed

Madagascar

Air Madagascar leased on 1049G freighter from Air France twice in 1965 for all-cargo service to Paris.

Rhodesia

Afro-Continental Airways used one 1049G from 1970 until 1974 on passenger and freight charters throughout southern Africa.

Air Trans-Africa operated one 1049G on charter work and in connection with the Biafran airlift from 1967 until 1970.

South Africa

Trek Airways leased two 749As from South African Airways in 1961 for nonscheduled low-fare services from South Africa to Europe. A ban on South African registered airplanes flying over certain African countries forced the return of the airplanes in 1963. Trek then entered a cooperative agreement with Luxair in 1964 to operate the flights.

Tunisia

Tunis Air operated one 749A and one 1049G briefly in 1961 on services to Paris and Zurich.

The Middle East

Israel

El Al became the only Constellation operator in the Middle East in May 1951, when it purchased three former C-69s for its Tel Aviv–New York service. Three 049s were added by December 1955, and all were converted to Model 149 standards with their interiors refurbished to 649 standards, along with the modifications needed for the 149.

Asia

Eight Asiatic countries had one Constellation operator apiece.

Cambodia

Royal Air Combodge used one Air France 1049G for services from Cambodia to Hong Kong from 1958 until replacing it in 1960.

Ceylon

Air Ceylon leased one 749A to restart Colombo (Sri Lanka)–London services in early 1956 in a first-class and tourist configuration. By 1958, a leased 1049G replaced the 749A and flew in a first-class and tourist/sleeper configuration. While Air Ceylon's airplane was in maintenance, it leased and used as many as four different 749s and six 1049s.

Japan

Japan Airlines leased a 1649 briefly from Air France in 1960 as a back-up aircraft on their joint Paris–Tokyo route.

Korea

Korean National Airlines (KNA) operated one 749A from 1959 until 1962 when KNA was reorganized as Korean Air Lines (KAL). Two 1049Hs were then leased for international services and used until 1967.

Malaysia

Malayan Airways leased QANTAS Super Constellations in 1960 for service to Hong Kong, but discontinued it shortly afterward.

Taiwan (Nationalist China)

China Airlines purchased one 1049H in 1966 in inaugurate scheduled international services from Taipei to Saigon, South Vietnam. The aircraft was later used for charters in Southeast Asia until it was retired in 1970.

Thailand

Thai Airways was actually a first-tier Connie operator, having ordered two 1049Cs in 1953. When the aircraft

This 1049G was delivered to KLM in May 1954 as 1049E PH-LKA. After conversion to 1049G it was sold to Air Ceylon in November 1958 and reregistered 4R-ACH. After a stay in Spain as EC-AQL, it regained its Dutch registration when repurchased by KLM in June 1962. ATP/Airliners America

A problem with identifying Chinese aircraft by their registration numbers is that both Nationalist China (Taiwan) and Mainland (Communist) China use B and numbers for their registrations. The 1049H shown as B-1809 belongs to China Airlines of Taiwan. ATP/Airliners America

The first of three 1049G-82s delivered to Thai Airways International in July 1957, HS-TCA was later sold to Mexico, Portugal, and Ireland before going to South Africa as a source of spare parts. Lockheed

A Model 1049G-82 of the Spanish airline Iberia on a factory test flight off the California coast before delivery to the airline on July 8, 1957. Lockheed

were never delivered, however, the airlines placed a new order for three 1049Gs. The airplanes were delivered by October 1957, and were used from Bangkok to Hong Kong, Tokyo, Rangoon, Burma, and Calcutta, India. But they proved to be too expensive to operate and were sold by 1958.

Vietnam

Air Vietnam leased 749s and 1049s in 1957 for weekly flights from Saigon to Hong Kong before replacing them three years later.

Constellations in Central and South America

From Mexico southward through all of South America, twelve nations supported thirty-two Constellation operators. It should be noted that the letters *S.A.* appearing after many of the company names do not stand for South America. They represent the term *Societe Anonyme*, meaning a stockholder company rather than a proprietorship or a corporation with limited liability (Ltd.).

Argentina

Aerolineas Carreras Transportes Aereos started using its 749A in July 1964 on freight charters to Miami. The carrier also served other points in South America until the airplane was impounded in Uruguay for smuggling in April 1967.

Aerotransportes Entre Rios SRL (AER) was formed in 1962 to provide domestic and international non-scheduled freight services. The airline soon specialized in moving horses and cattle with one 7490A, one 1049G, and three 1049Hs. The Super Constellations were retired by late 1970, but the 749A remained in service until 1972.

Aerovias Halcon SRL leased one 1649 freighter for six months in 1968, but made little use of it during that time.

Transcontinental S.A. received two 1049Hs in September 1958 for scheduled passenger services from Buenos Aires to New York and San Francisco. This service was offered three times a week using the 1049Hs. The flight included a stop in Rio de Janeiro. The airplanes were retired in March 1960.

Trans Atlantica Argentina acquired four 1649As in 1960 for a twice-weekly route from Buenos Aires to Geneva, Switzerland, with stops at Rio de Janeiro and Recife, Brazil, and Lisbon, Portugal. However, it became increas-

California Eastern received 1049H N468C in July 1958 and immediately leased it to Transcontinental S.A. of Argentina as LV-FTU. It was returned to Cal Eastern in January 1960. Damaged in a forced landing in Belize, it was scrapped. ATP/Airliners America

QANTAS received 1049E VH-EAA in February 1955 and converted it to a freighter five years later. After appearing on the US register as N9714C and in Canada as CF-PXX, it was acquired by Aerotransportes Entre Rios of Argentina as LV-IXZ in April 1967. ATP/Airliners America

One of four well-used 1649A-98s obtained by TransAtlantica Argentina S.A. from original owner TWA, seen here in its final days. Registrations were LV-GLH, GLI, HCD, and HCU. LV-GLI, repossessed by TWA, was used for an FAA crash test in September 1964. ATP/Airliners America

ingly unprofitable to operate these airplanes against what the competition introduced as jets, so on November 5, 1961, the airline suspended operations.

Barbados

Carib West used two 1049Hs briefly in 1974 and 1975.

Bolivia

Bolivian International Airways leased a 1049H from April until May 1975 for charter cargo flights.

Trans Bolivian bought one 749A in January 1968 but although the airplane was painted in Bolivian colors, it was never operated by the airline.

Transportes Aereos Benianos S.A. (TABSA) operated as a non-scheduled cargo operator with three leased aircraft—two 1049Hs and one 1049G—from July 1968 until suspending operations the following May.

Chile

Sociedad De Transportes Aeria LTDA used a leased 049 for cut-rate services from June 1957. A 1049G was used along with the 049 from late 1958 until the carrier ceased operations in 1959.

Dominican Republic

Aeromar leased an Air Cargo Support 1049H from 1977 until 1979 for freight flights to and from Miami.

Aerotours Dominicano operated one 1049 on passenger services within the Caribbean until a 1978 ban on carrying passengers in Constellations turned it into a freighter. One 1049G was added in 1979 before both aircraft were retired the next year and sold to Aerochago in 1981.

Aerovias Quisqueyana used three 049s and three 749s on scheduled services within the Dominican Republic and the Caribbean. Services were started in 1966 and continued until the 1978 ban on passenger-carrying Constellations forced Quisqueyana out of business. To its credit, the airline did operate the last passenger flight with a Constellation.

Argo S.A. purchased one 749A in April 1979 for use on cargo flights from the Dominican Republic to Miami and

KLM received 049-46-59 PH-TAW in June 1946. It was returned to Lockheed in October 1949, then used by Howard Hughes, TWA, and eleven subsequent lessors and owners before being acquired by Aerovias Quisqueyana of the Dominican Republic in December 1975 as HI-260. ATP/Airliners America

After retirement, in July 1968 VC-121A 48-614 went on the US civil register as N9466 and became a sprayer. Sold to a Miami dealer in January 1979, it was resold to Argo S.A. of the Dominican Republic in May 1979, and was registered HI-328. ATP/Airliners America

San Juan, Puerto Rico. Another 749 was purchased in November 1981, after the original airplane was lost in an accident in October, but the company ceased operations not long after it was delivered.

Transporte Aereo Dominicano purchased two 049s and one 749A in early 1979, but was out of business within a year.

Guatemala

Aviateca leased one 1049H from early 1972 until later in the year for cargo flights to Miami.

Haiti

Air Haiti purchased one 749A early in 1961 for services to the United States, but it was lost at sea later the same year.

Mexico

Aeronaves de Mexico leased two 049s in 1957 for services within Mexico before purchasing its own 749As in 1958. Another 749A was leased midyear following an accident involving one of the 749As, but all were displaced as front-line equipment by 1960.

Aerovias Guest S.A. can be regarded almost as a first-tier customer for the Constellation since its first 749, delivered June 6, 1947, had been registered previously only to Lockheed. Aerovias was the first airline owner. However, all of Aerovias' nine other Connies were obtained from other airlines.

Aerovias' first Connie, XA-GOQ, c/n 2503, was used on Mexico City–Miami flights and was leased briefly to Pan American. It was sold to Air France in January 1949, passed to the French Air Force as a test vehicle, and is now on display at Musee de l'Air at the Paris airport.

Three more 749s and three 749As were purchased in late 1955 following a name change to Guest Aerovias Mexico S.A. for services to Canada, the United States, and Panama. Three 1049Gs were purchased in late 1959 for services to Europe before all the remaining Connies were sold in March 1961.

Panama

Aero Fletes Internacionales S.A. (AFISA) used a 1049H on cargo flights from Miami to Panama City from 1970 until its permit was withdrawn in 1972.

Aerovias Panama started domestic services with Constellations in 1959 using two 1049Gs, but the airplanes were sold shortly afterward.

Rutas Aereo Panamenas S.A. (RAPSA) purchased one 1049G

After test by Lockheed as NX86520, the first 749-79-22 went to Aerovias Guest in Mexico as XA-GOQ in June 1947. Later, Guest leased it to Pan American, its parent company, and eventually sold it to Air France as F-BAZR. It is now in the Musee de l'Air in Paris as F-ZVMV. Lockheed

freighter in 1968, which was written off in an accident shortly afterward. Another 1049G was purchased and was flown on cargo flights for a short period before it was used in the Biafran airlift.

Paraguay

Lloyd Aereo Paraguayo S.A. (LAPSA) flew one 049 for passenger service from December 1963 until January 1964, when it was declared unairworthy and was abandoned.

Aerlinte Eireann acquired 749-79-32 EI-ACR in August 1947 and sold it to BOAC in May 1948 as G-ALAK. BOAC converted it to a 749A, then leased it to Skyways, which *converted it to a freighter. After registration in Uruguay as CX-BHD and Bolivia as CP-797, it went to COPISA in Peru as OB-R-899 in July 1968. ATP/Airliners America*

After return to Seaboard & Western, Aerlinte Eireann's 1049H-82 N1009C was sold to LEBCA of Venezuela as YV-C-LBP, shown here. It later acquired Bolivian reg- *istration CP-797 and Argentine registrations LV-PKW and LV-JJO. ATP/Airliners America*

Peru

Lineas Aereas Nacionalis S.A (LANSA) started low-fare service within Peru in 1964 with four 749As. Two more 749s were purchased in 1965, but after the crash of one LANSA suspended operations and never fully recovered.

Peru International operated one 749A on a weekly service to Miami starting in February 1966. The airline was reorganized in March 1967, and emerged as COPISA with an additional four 749As. The airplanes flew in a mixed passenger-cargo configuration until the last 749A was retired in December 1969.

Rutas Internacionales Peruanas S.A. (RIPSA) used one 749A on cargo flights to Miami. This carrier started operations in September 1966, but lost its operating certificate in May 1968 after a smuggling flight.

Trans-Peruana operated scheduled passenger services within Peru with four 749As from October 1967 until suspending operations during the summer of 1970.

Uruguay

Aerolineas Uruguayas purchased one 749A in January 1968 for nonscheduled cargo operations, but stopped all flights shortly afterward.

Compania Aeronautica Uruguaya S.A. (CAUSA) purchased two 749As in late 1962. A third was added in November 1963, and all were used on passenger flights within South America. A leased 1049H was added to the fleet in June 1966, but the carrier curtailed services the following May.

Rymar purchased an 049 in June 1965, but used it for smuggling flights so it was impounded by September of that year.

Venezuela

LEBCA owned two 1049Hs and occasionally leased another on freight services from Caracas to Miami starting in mid-1965. The company reorganized in mid-1966 after some financial difficulties and ultimately ceased operations in January 1968.

Venezolana Internacional de Aviacion S.A. (VIASA) used three 1049Gs for international services from April to December 1961.

Accidents and Incidents

The Constellation did not have an enviable safety record during its first two years of operation—eleven were lost in major accidents. But other than the troubles that resulted in the six-week grounding of July–August 1946, the causes were not attributable to any characteristics of the airplane itself. Many were the result of human error, while others were the result of mechanical malfunction or failure that could happen to any airliner design.

Pressurization Problems

A highly unusual accident occurred onboard a TWA 049 on March 11, 1947, between Gander, Newfoundland, and the Azore Islands as the navigator was taking a position fix through the plexiglass astrodome. It seems that as he was taking his readings, the astrodome broke and he was literally blown out of the airplane. It was the only astrodome incident to

occur on a Constellation and also resulted in the first airline casualty involving a pressurized cabin where a window or astrodome failed.

The exact cause of the accident was never determined, although it was believed to be an astrodome installation error.

The astrodome itself and its mounting were standard US Army items that had been used on the C-69 and many US transports and bombers during World War II, so the design was not at fault. To ensure that no future navigators would be lost, engineers installed a cumbersome body harness that was bolted to the cabin floor and was worn by the navigator while taking a celestial sighting.

A similar incident took place on April 20, 1957, when a passenger was sucked out a broken window on an Air France Super Constellation somewhere between Iran and Turkey, but that, too,

Believe it or not, Pan Am's 749-79-22, NC86527, flew the Atlantic with the number one propeller in this shape. It hit a snowbank while taxiing for takeoff from New York on January 4, 1948. No undue roughness was noted on the 16½ hour flight to London. Courtesy Chris Thornburg

During a landing with engine number three feathered, TWA 749-79-22 N91206 overran the runway in May 1948, but fortunately came to a stop before hitting the telephone poles and railroad tracks. R. Pavek via Norman E. Taylor

had happened on other pressurized airliners.

One of Pacific Northern Airline's 749s, N86523, experienced explosive decompression near Whittier, Alaska, on December 5, 1957. While climbing through 18,000ft, the forward left door blew out and portions entered the number three engine. This resulted in a loss of power, but the airplane was landed safely at Anchorage.

A somewhat humorous incident occurred when a female passenger became stuck in the lavatory of an 049. The flight engineer had been having trouble with the cabin pressurization system when a flight attendant came forward to the cockpit with the news. It was later determined that the fitting where the toilets are emptied and serviced had not been properly installed and had blown off in flight. The woman's bottom was serving as the stopper to keep cabin pressuriza-tion from being dumped overboard through the drainage system. The immediate fix was to depressurize the cabin in order to free her from the embarrassing situation.

Midair Collisions

Midair collisions, although not common, do occur occasionally with transport aircraft and the Constella-tion was involved in its share. The first involved a Pan Am 749 just after depar-ture from New York City en route to London. Another airplane, a small Cessna 170, was spotted by the Con-stellation crew just as it entered the wingspan of their plane. After pointing the nose sharply downward in an effort to avoid a collision, the 749 was struck just aft of the cockpit, where the Cessna ripped a hole 15ft long in the upper fuselage. No one on the Connie was hurt, but the two occupants of the Cessna were killed and the engine of their airplane remained jammed in the hole. A request for an emergency land-ing at a nearby Air Force base was made, and the airliner got down safely with only minor buckling of the skin.

Two of the most widely known cases of midair collision also involved Constellations. The first occurred over the Grand Canyon on June 30, 1956, when a United Airlines DC-7 struck TWA Super Constellation 1049-54-80 N6902C, *Star of the Seine*, from be-hind during good weather with only scattered clouds. Both aircraft crashed as a result of the collision and all onboard were lost. Because there were no witnesses to the actual collision and crash, the cause of the accident has never been fully determined. A second midair collision of TWA Super Constel-lation, N6907C, *Star of Sicily*, also involved a United airplane, this time a DC-8 jet. This occurred over Staten Island, New York, and once again the

Electrical fires were fairly common in early Constellations. Ground crews are fighting such a fire in this former C-69-5, 42-94559, now El Al's 049 4X-AKD, in Israel, Septem-ber 1953. Courtesy John Wegg

United plane collided with the TWA plane. It was later determined that the DC-8 was off course. All aboard both airplanes were killed.

Gunning for a Constellation

Even the smaller Constellation operators were not immune to error in their operations. A Pacific Northern 749 on a trip from Seattle to Kodiak, Alaska, became lost after encountering a 90 knot headwind. To counter the headwind, the crew dropped down to an altitude of 7,000ft but then encountered a static-producing ice cloud that rendered the automatic direction finders (ADF) inoperable. To try to regain the use of the ADFs, they climbed to a higher altitude, which also ensured that they would clear all potential mountains in the area. In a quest for signal beams, the Constellation made several turns before receiving the Kodiak ground-controlled approach (GCA) signal.

What the crew didn't know at the time was that two USAF fighters had been dispatched to help find the troubled airliner. The fighters made radar contact with an unknown airplane but watched as it made a course change and left the surveillance area. A second pass was made, but again the same thing happened as the Constellation was trying in vain to find a useful ADF signal. They came in for a third time and again the Constellation veered off on some unknown course. It seems that each time the plane changed course, it was at the same distance that an enemy pilot would pull away to avoid a rocket attack. Finally, the fourth time, with rockets armed, the pilots made visual contact with the Constellation. With positive identification made, the rockets were put back on "safe." Those inside the 749 knew nothing of the incident until later.

Oxygen Problem

Another serious problem surfaced on May 13, 1948, when TWA almost lost a Connie after the cockpit crew was knocked out by oxygen deficiency. In response to a false fire warning in the forward baggage compartment, the crew had discharged the CO_2 fire extinguisher. The odorless and colorless gas quickly filled the cockpit and nearly overcame the crew.

Lockheed recognized the problem and reported it to the industry via the Air Transport Association (ATA) and revised its extinguishing procedure to depressurize the cabin and open the cockpit windows before activating the CO_2. It also recommended the availability of 100 percent oxygen masks for the cockpit crew. Unfortunately, not all airlines took this warning seriously; another line lost a DC-6 and its passengers for exactly this reason.

Random Accidents

Some of the smaller air carriers, such as Imperial Airways and Paradise Airlines, were forced out of business after suffering major accidents with their Connies. Imperial was formed in June 1948 to provide charter services to the US government and began operating 049s. During Imperial's short life span it did not have an enviable safety record. One of its 049s was flying from Gander, Newfoundland, to Copenhagen, Denmark, on July 11, 1961, when an emergency landing had to be made at Prestwick, Scotland. This was because only two engines were operating and they had both blown cylinder heads. While on a military charter flight, another of Imperial's 049s was attempting to land at Richmond, Virginia, on November 8, 1961, but

crashed instead. This mishap resulted in the deaths of seventy-six servicemen, and the carrier's operating certificate was revoked within the month.

Paradise Airlines didn't even last that long. It was incorporated on June 27, 1962, but was out of business by March 1964. The company had provided charter services from California cities to Lake Tahoe using 049 Connies, but also suffered an accident. It occurred at Lake Tahoe on March 1, 1964, with the loss of all eighty-five passengers aboard. After the accident, the Civil Aeronautics Board (CAB) revoked the carrier's operating certificate, citing violations of safety regulations.

One of TWA's 049s, N686512, was lost on October 12, 1945, at Wilmington, Delaware, as the pilot attempted to land during a light rain with a wind shift. Braking action was poor and the airplane ran off the runway, through some trees, and across a road, hitting two cars. It eventually ended up 650ft from the runway, turned counterclockwise, and burst into flames.

The Venezuelan airline LAV's 1049E, YV-C-AMS, crashed in the Atlantic Ocean on June 30, 1956, off Asbury Park, New York. The airplane had taken off from New York on its way

Hawaiian airline Willair International's first 1649A, N8081H, was considered unfit for service after this landing gear accident at Stockton, California, on June 28, 1958, and was scrapped. ATP/Airliners America

to Caracas when the number two propeller started to overspeed and became uncontrollable. The airplane turned back to New York and had just begun dumping fuel when it exploded, with the loss of all onboard.

The first 1049H delivered to the Flying Tigers line, N6911C, was involved in an odd accident on December 1, 1961, at Grand Island, Nebraska. While the airplane was taxiing, the right main gear broke through the concrete and collapsed. After repairs, N6911C was destroyed in a landing accident at Adak Naval Air Station in the Aleutian Islands on March 15, 1962.

The largest loss of life on a single Constellation happened on March 15, 1962, when a Flying Tigers 1049H with 114 passengers, including crew, was lost at sea somewhere between Guam and the Philippines.

Mountain Crashes

Flying over mountains has always been one of the most dangerous duties facing transport pilots, and many airlines have suffered the loss of an airplane when it struck a mountain. One of Air India's 749 Connies struck Mont Blanc in France on November 3, 1950, with the loss of forty-eight passengers. Another airline that had to suffer such an indignity was Air France. The accident happened on September 1, 1953, when a 749 collided with Mont Cemet in France. Another mishap involved a Super Constellation belonging to LAV. It was lost on October 14, 1958, when the airplane collided with Mount Cedro in Venezuela with twenty-three passengers aboard. In the United States, Pacific Northern Airlines lost 749A-79, N1554V, to a mountain. This occurred

Because of frequent in-flight engine failures, Connies were often called The World's Best Trimotors, but here is the only true trimotor Connie. After Pan American's 049-46-26 had engine number four fall off, the remains of the nacelle were removed, the gap in the wing was faired over, and the Connie was flown back to the factory for repair. Lockheed

on June 14, 1960, when the airplane unexplicably hit Mount Gilbert in Alaska with the loss of fourteen people.

Pan Am lost one of its 049s on June 22, 1951, when it crashed into a hill fifty miles from Monrovia, Liberia, while en route from Johannesburg, South Africa, to New York. The airplane was flying through very bad weather and heavy rain when it descended below the minimum en-route altitude with the loss of all those onboard.

Military Mishaps

The military also had its share of incidents with the Constellation. One of the first C-69s produced was forced to make a landing in the Mediterranean in 1945, but was salvaged and sold to the French government. On October 30, 1954, an R7V-1 disappeared while flying from Maryland to North Africa with forty-two persons aboard. Although an extensive search was conducted, nothing from the airplane was ever found. A C-121C was destroyed on December 30, 1956, when it struck the ground short of the runway while landing in heavy fog. The airplane was on a routine MATS flight at the time of the accident, and fifteen out of the forty-one onboard were killed.

The USAF lost an RC-121D on July 11, 1965, after it ditched in the Atlantic Ocean northeast of Nantucket Island. There were three survivors. Initially, the number two engine had failed to shift to high blower and was left in low blower. Later the number three engine caught on fire, which was followed by the failure of the number two engine.

Military pilots are not immune to making mistakes, such as the time when an R7V-1 received severe damage when the landing gear was accidentally raised while landing. This was neither the first nor the last time that a copilot, either civil or military, accidentally raised the landing gear instead of the flaps after the airplane landed.

A WV-2 was lost on October 31, 1960, while landing in Antarctica after a flight from New Zealand. The airplane hit a snowbank on landing, sheared off the port wing, and broke the fuselage into several pieces. To their credit, none of the C-121As were ever involved in an accident before the last was retired on August 23, 1968.

On a humorous note, an incident occurred during the Vietnam War when an EC-121D was returning from a refueling stop in Da Nang. After cruising at 8,000ft on their way back to station, the pilots started their descent through broken clouds. When they broke out of the clouds, they found a very large aircraft carrier right in front of them. They were on a perfect approach and on glide slope when the landing signal officer began frantically waving them off. The crew, stunned by seeing a carrier in front of them, pulled up and missed the bridge by several feet.

Three-Engine Constellation

One of the most unusual flights of a Constellation was the one in which the airplane was fitted with only three engines. In June 1946, Pan American's 049 NC88858 encountered trouble with the number four engine shortly after takeoff for a transatlantic flight. The trouble resulted in the engine falling completely off of the airplane, after which the Connie was landed safely.

After inspection, it was decided that repairs could be made more conveniently at the factory than at the remote East Coast field. It was suggested that the rest of the number four engine nacelle be removed, the open leading edge of the wing be covered over with sheet aluminum, and the airplane be ferried back to the factory. The FAA approved the scheme and the trimotor NC88858 made a transcontinental flight at reduced gross weight.

Biafran Airlift

Of course, not all Constellation incidents were in the negative column; for example, the Biafran airlift that was conducted from May 1967 until January 1970 was a positive event. Biafra was the southeastern region of Nigeria in West Africa and also the main oil-producing part of the country. When more than 20,000 civilians were slaughtered during the unrest in that region, the leader of the local tribe declared independence from the government. Civil war erupted almost immediately, and Biafra was soon almost completely dominated by the Nigerian military. To provide assistance to the new country, many organizations were

started with the intent of supplying food, ammunition, and needed medical supplies. Over ten different organizations participated in the airlift, including the Biafran government. The Constellation proved to be an ideal airplane for this role, and the models used included 749s, 1049s, and one 1649. Of the twenty-four Connies participating in the relief effort, five were lost in landing accidents and another two were destroyed on the ground.

Eastern Shuttle

An equally astounding, but not as dangerous, event occurred on April 30, 1961, when Eastern Air Lines inaugurated its now-famous Air Shuttle among Washington, New York, and Boston. Originally only eight ninety-five-passenger 1049Cs were used on the shuttle, but within the next two years all thirty-two of the remaining Connies, comprised of 1049, 1049C, and 1049G models, were in use, plus some Douglas DC-7Bs. This was the first no-reservation shuttle service that guaranteed a seat for anyone purchasing a ticket. It was not uncommon for an extra flight to be dispatched with only one passenger aboard. Service was initially provided from 8:00am to 10:00pm on the even hours but was later increased from 7:00am to 10:00pm, departing every hour. Eventually, the Connies were displaced from first-line service and used only as back-up aircraft until February 14, 1968, when the shuttle's last flight took place.

Pacific Northern Adventures

There are countless incidents involving Connies of the various airlines that would make interesting reading, but coauthor Peter Bowers has picked three from Seattle-based Pacific Northern Airlines (PNA) of which he has direct information.

The Fire That Wasn't

Spotting a southbound flight from Alaska to Seattle, a ground observer at the north end of Vancouver Island (Canada) reported to the PNA Connie by radio that it was trailing black smoke. A little later, another observer farther south on the island reported the same thing. The crew checked the airplane as best they could in flight, but found no evidence of a fire. How-

The Eastern Air Lines Air Shuttle from Boston, Massachusetts, to Washington, D.C., which operated from 1961 into 1968, was a revolutionary bit of airline business. This 1049-53-67, N6213C, was bought by Eastern in February 1962 and leased to Pan Am in 1955. It ended its airline career on the Eastern Shuttle. ATP/Airliners America

ever, they thought that they should mention the reports to their destination at Seattle's Seattle–Tacoma (SeaTac) airport. Taking no chances, the PNA staff there alerted the airport emergency crews and fire engines were spotted along the runway in anticipation of a fire onboard the soon-to-arrive Connie.

Now things got scary. The Connie reported over northern Seattle and was cleared by the tower for a straight-in landing at SeaTac. In those preenvironmental awareness days, there was a large dump at the south Seattle city limit that was located right under the flight path into SeaTac. Just about the time the PNA Connie was expected to be visible from the airport, someone touched off a fire at the dump that put up a huge cloud of black oily smoke, just the kind that would result from a large airplane crash. Those scanning the sky from the tower and the ground at SeaTac groaned, figuring that the

Connie had just crashed. Gloom turned to joy in a moment, however, as the Connie's silver nose popped out of the smoke and the plane soon rolled to a smooth landing, trailed by all the airport's fire equipment.

What had happened? As well as could be determined, the observers had seen the shadow of the Connie's vapor trails on a thin undercast. Their angle of sight relative to the sun and the airplane made it look like the dark shadows were coming directly from the airplane.

Emergency Communications

The major earthquake that rocked Fairbanks, Alaska, on March 27, 1963, knocked out all communication from the city to the outside world for a while. The first radio communication from Fairbanks to outside was by means of the equipment aboard a PNA Connie parked near the airport tower and operated by the same pilot who had had the nonfire adventure.

Unofficial Passengers

Pacific Northern's flights to Alaska originated at SeaTac, but for a while the airline's maintenance base was at Seattle's Boeing Field, some six miles to the north. This meant that after discharging its passengers at SeaTac, the airplane had to be ferried over to Boeing Field. If the wind was from the north, this was a very short flight—off the ground, jog slightly to the right, and make a straight-in landing at Boeing. If, however, the wind was from the south, the takeoff was to the south and was followed by a roundabout northern route that would bring the plane into Boeing Field from the north. When such a flight was to be made, some PNA pilots would invite their friends along for a sightseeing tour of Puget Sound. At one time, a whole Boy Scout troop with which this coauthor was associated rode along.

Flying and Fadeout

This chapter brings us to the end of the Lockheed Constellation's active life as an airliner and military transport-patrol plane. Throughout its service, it was affectionately known as the Connie and was—and still is—regarded as the most beautiful piston-engine transport ever built.

Production was started as an airliner in 1942 and ended, still as an airliner, in 1958, sixteen years later. During those years, the US armed forces bought 331 of the 856 Connies built.

Over those sixteen years the Connie grew from an 82,000lb airplane with 2,200hp engines to a 160,000lb plane with 3,400hp engines. It grew in size, too, from an original wingspan of 123ft and a length of 95ft, 2in, to a span of 150ft and a length of 116ft, 2in.

The Connies were forcibly retired from the airlines long before they wore out, by the introduction of jets, but many were able to carry on, after modification, as sprayers and freighters. The military Connies lasted longer, with the last US Navy NC-121K, formerly a WV-2, retiring in June 1982. Some retired military Connies found new civil owners who were able to put them to work again.

Only a few Constellations are known to be flyable in the United States today, but many others are on static display at various museums and

The last scheduled Connie passenger flight by a US airline was with this Western Airlines 749A-79 from Juneau, Alaska, to Fairbanks on November 26, 1968. It had been built for Air India as 749-79-22 VT-CQR in February 1948, passed to QANTAS as VH-EAE, which converted it to a 749A-79, then went to BOAC as G-ANTG before going to Pacific Northern as N1552V in November 1958. Pacific Northern and its six surviving Connies were absorbed by Western Airlines in July 1967. Museum of Flight

military air bases (see chapter 16). The flyable articles are maintained by their enthusiastic owners as flying museums, dedicated to preserving the sight and sound of the Connie for many years to come.

Flying the Constellation

During the immediate postwar years and throughout the 1950s and early 1960s, Lockheed and Douglas battled it out for airliner supremacy, the former with the Constellation in its various models and the latter with the DC-4, DC-6, and DC-7 series. From the beginning Lockheed with its 049 Connies offered a larger aircraft with better range than the competing DC-4, which was an older design. Over the next fifteen years, Douglas slowly closed the performance and size gap, although its last DC-7C model still grossed some 17,000lb less than the final Connie, the Lockheed 1649A. Trade-offs in performance were fairly evenly split. The DC-7C offered better cruise, and was cheaper by nearly $500,000. The 1649 countered with 300 miles greater range and more payload.

As for the relative merits of their flying qualities, Merlin Fleming, a former military ferry pilot and long-time commercial pilot with more than 30,000 hours, claims that the Douglas airliners were easier to fly and more responsive to the pilot's input. He liked the Connie, but said it was heavier on the controls, more like a truck, mainly because it was a bigger airplane. Long-range Connie flights were more work because every time the superchargers went into high blower as the aircraft climbed, the danger of overstraining the engines and dramatically increasing the chance of engine failure and fire increased. Consequently, Fleming learned to pull back on the power and cease climbing every time the superchargers went into high blower. It took longer to reach cruise altitude that way, but it relieved the strain on the engines and once some of the fuel had burned off, he was eventually able to climb in increments of a few thousand feet in low blower.

Noted Lockheed test pilot Herman "Fish" Salmon was killed in the crash of an old and tired Constellation he was having renovated at Indianapolis. Among a number of revisions he was making to the 1049 in the mid-1980s was the replacement of its hydraulic props with electrically actuated ones. The aircraft lost power on takeoff and Salmon, who was flying in the copilot's seat, died as the Connie veered off the runway. Salmon had enjoyed an illustrious career as a Lockheed test pilot and exhibited the scars on his body from several bailouts and accidents. Not generally known about was his clandestine tour of Vietnam with Lockheed's then super-secret Cheyenne attack helicopter, which Salmon flew for a month on combat operations.

Final Operations

TWA's phaseout of its Constellation fleet is typical of other major airlines' operations. Its last transatlantic 1049A flight was on August 28, 1961, and domestic passenger flights ended two years later, with the last one retired in 1965. The 1049Gs lasted longer, until 1966, with one serving into 1967. The last Connies in TWA service were 749As and the 1049Gs. A 749A made TWA's last passenger flight on April 6, 1967, and a 1049G carried the last load of freight on May 11. From that day on, TWA became an all-jet airline. The last scheduled passenger Connie flight on a US airline was by Western Airlines with a 749A from Juneau, Alaska, to Fairbanks on November 26, 1968.

Those airline Connies that were not scrapped by their original owners passed to smaller airlines and to independent operators at whose hands they endured considerably less than airline standards of maintenance and operation. Some, including surplus military models, were converted to spray planes, and one was tested as a tanker for fighting forest fires, but the Forest Service did not accept it because it required a three-person cockpit crew. The few Connies still in significant commercial operation in 1991 were based in the Republic of Santo Domingo. One of these came to a spectacular end on April 5, 1990, when an in-flight fire dropped an engine off the plane and it ditched in the Caribbean off Puerto Rico. Sequence photos of this ditching made the front pages of US newspapers on a slow-news day.

There are now less than half a dozen flyable Connies left in the United States but they are private "flying museum" projects; pampered exhibition items, not workhorses. Their owners think enough of the Connie and its significance to air transportation to go to great lengths to preserve these few airworthy survivors.

Surviving Constellations

As with any airplane that has been out of production for over thirty years, few models of the Constellation remain. Most became redundant just after being retired from first-line service; others have continued to soldier on in roles never intended for them. Of the original 856 built, few remain today and this number is steadily dwindling. Most are in anything but airworthy condition and will likely end up being sold for scrap or used for spare parts to keep their brothers flying. Sporadic attempts have been made to restore some of the remaining Constellations to airworthy condition, but few have been successful.

Survivors by the Numbers

Exact figures are not available, but seventy-four Constellations are known to survive today. They are listed in sequence of Lockheed serial number, civil registration or military serial number, location, and status in this chapter. General and detailed discussions of Constellation survivors follow.

The short-fuselage Constellations are represented by nine airplanes—four 049s, one 649, and four 749s. The Super Constellation series is faring a little better, with twenty-one models of the type still in existence. These include one 1049, three 1049Cs, one 1049E, nine 1049Gs, and seven 1049Hs. The number of military Constellations remaining, at forty-four, is larger than the combined total of all the civilian models. These military models include one C-69, seven C-121s, one PO-1, six R7V-1s, one VC-121E, eight C-121Cs, six WV-2s, and eight RC-121Ds. Six Starliners also remain. These numbers are for the original production models and do not include any upgraded or modified airplanes.

At the time of this writing, some of the remaining Constellations are just

Still in its former Capitol Airways colors, this 1049E-55, formerly CU-P-573 but now N1005C, sits proudly in its elevated position in downtown Penndale, Pennsylvania. Note the spiral staircase and absence of engines and propellers. This Connie was the inspiration for the Greenwood Lake 049. "Passengers" were served dinner in this annex to the restaurant below. Curtis K. Stringfellow

Originally purchased by Air France on November 4, 1953, this 1049C F-BRAD also saw service during the Biafran airlift. It has been stored at Nantes, France, since May 1974 and is a local tourist attraction. This picture shows the airplane in the colors of the original 1049 prototype, but it has recently been repainted in Air France colors. ATP/Airliners America

This 749-79-33, originally KLM PH-TET and later converted to 749A, had been rotting at Dublin airport since 1974 before it was purchased by Aces High in 1982. Initial plans were to restore it to flying condition, but it ended up being dismantled and shipped to the United States as deck cargo. After many owners, it was then rebuilt and painted to resemble a 1960s era TWA 749 with its first US registration N7777G. The plan is to use it for demonstration purposes and films. ATP/ Airliners America

waiting for the cutter's torch, with no chance of ever being flown again. One such airplane, an 049, sits at Fort Lauderdale without props but has been purchased by a new owner, although no work has started. Some of the Super Constellations used during the Biafran airlift are still at Sao Tome, West Africa, and are likely to remain there. Another 1049H freighter sits at Miami International airport and is scheduled to be used as a fire service trainer while another, at Opa Locka, Florida, is currently being dismantled. A former TWA 1649 freighter has been at Paramaribo, Suriname, since landing there with engine trouble during a smuggling flight. In northern Chile another 1649 crashed on a mountain outside Isluga and hasn't been moved since.

Aerochago S.A. of Santo Domingo currently operates the largest fleet of Constellations in the world with four (one 749, one 1049, one R7V-1, and one C-121C). A C-121A is about to be added to the fleet. The last flying 1049 has seen better days; this 749, HI-422, is missing two of its engines and will probably never fly again. ATP/Airliners America

126

Constellations on Display

This all sounds disheartening until taking a closer look at the number of Constellations currently on display. By far the most enjoyable is the C-69 at the Pima Air Museum in Tucson, splendidly repainted in TWA colors. This is the oldest Constellation in existence and is also the only remaining C-69. Undoubtedly the most striking Constellation of all is the 049 at Asuncion, Paraguay, which has been painted in an eye-catching pop-art scheme. Another 049 is in Santa Cruz, Bolivia, corroding away in a bare-metal finish inside a children's playground.

The 049 in Greenwood Lake, New Jersey, has been turned into a cocktail lounge but was opened to the public for only a brief period in 1981. Just out-

This RC-121D, 52-3418, served with the Air Force Reserve's 79 AEW & CS at Homestead, Florida, until April 7, 1976. Although said to have been converted to an EC-121H, it is believed that it was only *redesignated as an EC-121D before being converted to an EC-121T. The Combat Air Museum at Topeka, Kansas, purchased the airplane in March 1981 and currently has it on display.* John Campbell

Aerolineas Mundo S.A., or AMSA, operates one C-121C, HI-515, out of Santo Domingo. A 1049H that had been stored at Sanford, *Florida, for many years is about to be added to the fleet.* ATP/Airliners America

After President Eisenhower's VC-121A 48-610 Columbine II *was retired in 1968, it was sold to a spray operator as a source of* spare parts. No one looking at this photo, taken after nearly twenty years of inactivity in a desert boneyard, would ever *expect that the plane would fly again.* Alan Manley

Miracles of restoration do happen. Here is Columbine II, *restored to its presidential* glory, *after its first flight in twenty-two years, April 5, 1990.* Alan Manley

side Philadelphia, a 1049 sits high atop a restaurant on airfoil-shaped supports. In Europe, perfect examples of airline Super Constellations are on display in Frankfurt and Hermeskell, West Germany, with another two in Quimper and Nantes, France. The Wroughton Science Museum in England has painstakingly restored a 749 with 1960-era TWA markings. The only Starliner currently on display is the prototype in Nagoya, Japan, which serves as a restaurant there.

The US military has recently taken some interest in the Constellation's history as more of the old early-warning and transport Constellations are flown to and displaced at their original bases. At Fort Rucker, Alabama, an excellent example of a C-121 has been preserved in NASA markings. This airplane also served in the Korean War as the personal transport of General MacArthur. Numerous early-warning Constellations have recently been saved and relocated to such areas as California, Georgia, South Carolina, and Texas. President Eisenhower's personal transport during his presidency, *Columbine III*, is on display at the US Air Force Museum in Dayton, Ohio, along with an early-warning version. Still more military Constellations are scheduled to leave the storage area at Davis-Monthan AFB, Arizona, in the near future so there won't be any shortage of display models in the United States.

Restoring Organizations

Even better news is the newly stimulated interest in the Constellation itself. A number of organizations have purchased Constellations in the hope of making them flyable once again. One of the first, and so far the most successful, is Save-A-Connie (SAC). This nonprofit organization, based in Kansas City, Missouri, has beautifully restored a 1049H into a Super-G and in doing so has produced the only civil Constellation now in the world with wing-tip tanks. Although it was originally intended that the airplane would carry TWA markings, insurance regulations dictated that SAC be used on the airplane in the appropriate areas. The interior is scheduled to be completed to full airline standards with a seventy-passenger interior along with the appropriate interior fittings. SAC plans to

TWA's Starliner 1649A-98 N7306C did not go to another operator after the airline retired it in April 1962. After three years in storage, it became the Flight 42 Cocktail Lounge in Kansas City, Missouri, but did not last long as such. Note the misaligned propeller blades, a common fault of many display aircraft. Don C. Wigton via Ed Peck

fly its Constellation to air shows around the United States and should be doing so for a long time.

Some other organizations trying to restore Constellations are South African Airways, the Dutch Association of Constellation Enthusiasts, and Starliner Promotions. South African Airways recently completed the restoration of a Starliner for static display after the plane had sat deteriorating for several years in open storage. The airplane arrived in Johannesburg the summer of 1979, but restoration work didn't start until almost five years later, in 1984. The Dutch group was founded in early 1988 to restore a Constellation in KLM colors for European air shows. A suitable 749 was chosen, a former Conifair sprayer in storage in Canada, and the group is continuing to raise the money needed to purchase the airplane. Once the airplane arrives in Holland it will be restored, at least externally, and will be visiting air shows for two years before being put on permanent display. The last group currently trying its hand at Constellation restoration is Maine Coast Airways. Under the name Starliner Promotions, the group is trying to restore a 1649 to its original Lufthansa colors along with a full interior.

Even film star John Travolta has tried his hand at restoring one of these magnificent airplanes. He also purchased one of the former Conifair 749s in 1984, with the intent of restoring it to represent the typical Constellation of the 1950s. The airplane was first flown to California but was later transferred to Arizona, where it stayed basking in

the sun untouched for some time. His interest in the airplane eventually waned and it was sold to a group of airline enthusiasts in mid-1987. This group plans to restore the airplane, using the forward part of the interior to depict a standard airline and the rear to house a museum of all Constellations in airline service. If all goes as planned, the airplane should be delighting air show crowds for many years to come.

There is also talk of restoring an 049 series Constellation. The airplane in question is the former Air France and TWA 049 currently at Greenwood Lake, New Jersey. It seems that a museum is trying to persuade the owners of the airplane to donate it to them for restoration. Plans call for having the airplane ready to ferry within a year and on the air show circuit in another two.

Flyable Connies

Nine Constellations are still flying at this writing. Most of these are in the Dominican Republic and are operated from Santo Domingo to Miami and other Caribbean destinations. Opera-

Next page

Of the 856 Lockheed Constellations built, very few are still flying in 1992. The most original airline version is this 1049H, N6937C, originally delivered to Slick Airways in September 1959. Derelict since 1975, it has been carefully restored over a four-year period by an organization named Save-A-Connie (SAC), and first flew in its present SAC markings in 1990. Arnold Swanberg

tors include Aerochago S.A. and Aerolineas Mundo S.A., or AMSA.

Besides the Save-A-Connie 1049H, the only Constellations flying in the United States are two owned by Maine Coast Airways of Auburn, Maine. These 1649 Starliners are the only ones of their type still flyable.

Another Super Constellation that is registered in the United States be-

longs to World Fish and Agriculture, Inc. Aptly named *Winky's Fish*, the Constellation is stationed on the tiny island of Belau, formerly Palau, just east of the Philippines, to make twice-weekly runs to Nagoya, Japan, with fresh tuna. Unfortunately, just after starting operations, the airplane was impounded at Manila after falling behind in its scheduled flights. It now

sits alone on an apron there and will probably never fly again.

One of the most eagerly awaited operations was that of Classic Airlines at Van Nuys, California. It had planned to fly charters with two restored Super Constellations and had one with a complete interior, but this unfortunately came to nothing even though official approval had been granted.

This table lists all of the Lockheed Constellations known to exist worldwide in 1991. They are arranged in sequence of manufacture, which explains why the low serial numbers in column two are at the end of the list—the Model 1649A Constellations, the last ones built, had lower serial numbers than any other Constellations. The first column on the left identifies each airplane by its last active civil registration or military serial number, or the one currently applied to a display airplane. The second column lists the constructor's number, and the third column gives the civil or military designation of the airplane. Columns four and five give the location and present status. A question mark (?) means that the authors are unsure of the present status of that particular Constellation.

Constellation Survivors

Registration	C/N	Type	Location	Remarks
N90831	1970	C-69	Tucson, AZ	Pima Air Museum
N86533	2071	L-049	Asuncion, Paraguay	Displayed
N9412H	2072	L-049	Greenwood Lake, NJ	Restoration (?)

Registration	C/N	Type	Location	Remarks
N90816	2078	L-049	Ft. Lauderdale, FL	Scrapping (?)
N2520B	2081	L-049	Santa Cruz, Bolivia	Displayed
F-ZVMV	2503	L-749	Paris, France	Musee de l'Air, Le Bourget
HI-332	2522	L-649	Agua Jira, Colombia	Impounded (?)
N7777G	2553	L-749	Wroughton, United Kingdom	Science Museum/TWA
N608AS	2600	VC-121B	Tucson, AZ	Stored
N494TW	2601	C-121A	Tucson, AZ	Stored/Future Restoration
48-0610	2602	C-121A	Davis-Monthan AFB, AZ	Stored
HI-393	2603	C-121A	Santo Domingo, Dominican Republic	Flying/ Aerochago
C-GXKR	2604	C-121A	St. Jean, Quebec	Dutch Connie Society
N422NA	2605	C-121A	Ft. Rucker, AL	Preserved/ Bataan
48-0614	2606	C-121A	Tucson, AZ	Flying/ Columbine II
N1206	2613	PO-1W	Salina, KS	Stored/FAA
HI-422	2667	L-749A	Santo Domingo, Dominican Republic	Flying/ Aerochago
CN-CCN	2675	L-749A	Casablanca, Morocco	Trainer
HI-228	4009	L-1049	Santo Domingo, Dominican Republic	Stored/ Scrapped (?)
54-4052	4106	R7V-1	Davis-Monthan AFB, AZ	Stored
54-4062	4137	R7V-1	Davis-Monthan AFB, AZ	Stored (?)
N420NA	4143	R7V-1	Aberdeen Proving Ground, MD	Stored
N4247K	4144	R7V-1	Manila, Philippines	Impounded
53-7885	4151	VC-121E	Wright Patterson AFB, OH	USAF Museum/ Columbine III
HI-532CT	4155	R7V-1	Santo Domingo, Dominican Republic	Flying/ Aerochago
131655	4156	R7V-1	Davis-Monthan AFB, AZ	Hulk (?)
54-0155	4174	C-121C	Lackland AFB, TX	Preserved
N73544	4175	C-121C	Camarillo, CA	Stored
54-0157	4176	C-121C	Davis-Monthan AFB, AZ	Stored
54-167	4186	C-121C	Davis-Monthan AFB, AZ	Stored (?)
HI-515	4192	C-121C	Santo Domingo, Dominican Republic	Flying/AMSA
N1104W	4196	C-121C	Dulles Airport, DC	Smithsonian Collection
54-0180	4199	C-121C	Charleston AFB, SC	Preserved/ MATS Atlantic
HI-548	4202	C-121C	Santo Domingo, Dominican Republic	Flying/ Aerochago
128324	4304	WV-2	Davis-Monthan AFB, AZ	Hulk (?)
N4257L	4335	RC-121D	Helena, MT	Preserved
N4257U	4336	RC-121D	Topeka, KS	Combat Air Museum
52-3425	4343	RC-121D	Peterson AFB, CO	Displayed
137890	4347	WV-2	Tinker AFB, OK	Preserved
53-0535	4350	RC-121D	Davis-Monthan AFB, AZ	Stored
53-0548	4363	RC-121D	Tucson, AZ	Pima Air Museum
53-0554	4369	RC-121D	Tucson, AZ	Pima Air Museum
53-0555	4370	RC-121D	Wright Patterson AFB, OH	USAF Museum
141292	4416	WV-2	Florence, SC	Air & Missile Museum
141297	4421	WV-2	Warner Robbins AFB, GA	Preserved
141309	4433	WV-2	McClellan AFB, CA	Preserved
141311	4435	RC-121D	Chanute AFB, IL	Preserved
143221	4495	WV-2	Pensacola, FL	Navy Museum
F-BRAD	4519	L-1049C	Nantes, France	Displayed/Air France
HI-329	4536	L-1049C	Santo Domingo, Dominican Republic	Derelic
CF-RNR	4544	L-1049C	St. Jean, Quebec	Preserved
N1005C	4557	L-1049E	Penndale, PA	Displayed/ Restoration (?)
D-ALIN	4604	L-1049G	Hermeskell, W. Germany	Displayed/ Museum
5N-83H	4616	L-1049G	Albufeira, Portugal	Preserved
5T-TAF	4618	L-1049G	Kirkop, Malta	Preserved
F-BHBG	4626	L-1049G	Quimper, France	Displayed
IN-316	4666	L-1049G	Bombay, India	Stored (?)
F-BHML	4671	L-1049G	Frankfurt, Germany	Int. Airport/ D-ALAP
VP-WAW	4685	L-1049G	Harare, Zimbabwe	Preserved
BG-583	4686	L-1049G	Pune, India	Air Force Museum
BG-579	4687	L-1049G	Pune, India	Stored (?)
N1007C	4805	L-1049H	Opa Locka, FL	Dismantled
N1880	4820	L-1049H	Miami, FL	Derelict/Fire Training
HI-542CT	4825	L-1049H	Santo Domingo, Dominican Republic	Flying/AMSA
N6937C	4830	L-1049H	Kansas City, MO	Flying/Save-A-Connie
CF-NAL	4831	L-1049H	Sao Tome, W. Africa	Derelict (?)
CF-NAM	4832	L-1049H	Sao Tome, W. Africa	Derelict (?)
N469C	4847	L-1049H	Sebring, FL	Ground Training
N1102	1001	L-1649A	Nagoya, Japan	Preserved
N7311C	1013	L-1649A	Isluga, Chile	Abandoned
N7315C	1017	L-1649A	Anchorage, AK	Derelict
N7316C	1018	L-1649A	Auburn, ME	Flying/Maine Coast
N7319C	1030	L-1649A	Paramaribo, Suriname	Derelict (?)
N8083H	1038	L-1649A	Auburn, ME	Flying/Maine Coast
N974R	1040	L-1649A	Sanford, FL	Flying/Starliner Promotions
ZS-DVJ	1042	L-1649A	Johannesburg, South Africa	Preserved

Constellation Specifications

	049	649	749A	1049
Wingspan	123ft	123ft	123ft	123ft
Length	95ft, 3in	95ft, 3in	97ft, 4in *	113ft, 7in
Height	23ft, 8in	23ft, 8in	22ft, 5in	24ft, 9.4in
Cabin length	64ft, 9in	64ft, 9in	64ft, 9in	83ft, 2in
Cabin width	10ft, 8.6in	10ft, 8.6in	10ft, 8.6in	10ft, 8.6in
Cabin height	6ft, 6in	6ft, 6in	6ft, 6in	6ft, 6in
Empty weight	49,392lb	55,000lb	56,590lb	69,000lb
Max takeoff weight	86,250lb	94,000lb	107,000lb	120,000lb
Max landing weight	75,000lb	84,500lb	89,500lb	101,500lb
Max payload	18,423lb	20,276lb	20,276lb	19,335lb
Max speed	339mph	352mph	345mph	
Cruise speed	313mph	327mph		301mph
Range (max fuel)	3,995 miles	3,995 miles	4,845 miles	
Range (max payload)	2,290 miles		2,600 miles	2,610 miles
Service ceiling	25,300ft	25,700ft	24,100ft	25,700ft

	1049C	1049G	1049H	1649A
Wingspan	123ft	123ft, 5in (†)	123ft, 5in (†)	150ft
Length	113ft, 7in	116ft, 2in (*)	116ft, 2in (*)	116ft, 2in (*)
Height	24ft, 9.4in	24ft, 9.4in	24ft, 9.4in	23ft, 4.8in
Cabin length	83ft, 2in	83ft, 2in	83ft, 2in	83ft, 2in
Cabin width	10ft, 8.6in	10ft, 8.6in	10ft, 8.6in	10ft, 8.6in
Cabin height	6ft, 6in	6ft, 6in	6ft, 6in	6ft, 6in

	1049C	1049G	1049H	1649A
Empty weight	70,083lb	73,016lb	69,326lb	85,262lb
Max takeoff weight	133,000lb	137,500lb	137,500lb	156,000lb
Max landing weight	110,000lb	113,000lb	113,000lb	123,000lb
Max payload	26,400lb	24,293lb	40,203lb	24,355lb
Max speed	374mph	366mph	366mph	376mph
Cruise speed	314mph	311mph	311mph	342mph
Range (max fuel)	4,760 miles	4,815 miles	3,463 miles	6,885 miles
Range (max payload)	2,880 miles	4,165 miles	1,890 miles	5,410 miles
Service ceiling	23,200ft	22,800ft	22,800ft	23,700ft

	C-121A	R7V-1	C-121C	RC-121D/WV-2
Wingspan	123ft	123ft	123ft	123ft
Length	95ft, 3in	116ft, 2in	116ft, 2in	116ft, 2in
Height	24ft, 9in	24ft, 9in	27ft	27ft
Empty weight	61,324lb	70,000lb	72,815lb	80,611lb
Max takeoff weight	100,520lb	135,400lb	137,500lb	143,600lb
Max speed	334mph	368mph	368mph	321mph
Cruise speed	324mph	355mph	355mph	301mph
Combat radius	1,150 miles	1,150 miles	1,150 miles	1,150 miles
Combat range	2,385 miles	2,085 miles	2,085 miles	4,620 miles
Service ceiling	24,400ft	22,300ft	22,300ft	20,600ft

* With radar fitted.
† With wing-tip fuel tanks.

Major Operators

Even though 856 Constellations were produced, there are only about 300 known owners or operators of the type, including the military. This works out to just under three airplanes per operator, about the average fleet, although some operated only a single airplane. All operators that operated ten or more Constellations are listed in descending order of quantity.

- United States Navy: 205 total, including two PO-1s, fifty-two R7V-1s, two R7V-2s, 141 WV-2s, and eight WV-3s.
- TWA: 156 total, including forty 049s, forty 749s, ten 1049s, twenty-eight 1049Gs, nine 1049Hs, and twenty-nine 1649s.
- United States Air Force: 142 total, including fifteen C-69s, ten C-121As, thirty-two C-121Cs, ten RC-121Cs, seventy-two RC-121Ds, one VC-121E, and two YC-121Fs.
- Eastern Air Lines: seventy-nine total, including twelve 049s, fourteen 649s, seven 749s, fourteen 1049s, sixteen 1049Cs, one 1049E, ten 1049Gs, and five 1049Hs.
- Air France: sixty-two total, including four 049s, twenty-four 749s, ten 1049Cs, fourteen 1049Gs, and ten 1649As.
- KLM: forty-six total, including four 049s, twenty 749s, nine 1049Cs, four 1049Es, six 1049Gs, and three 1049Hs.

- Pan American World Airways: thirty-four total, including twenty-nine 049s, four 749s, and one 1049.
- Pennsylvania Air National Guard: thirty-one total, including fourteen C-121Cs, thirteen C-121Gs, and four EC-121Ss.
- BOAC: twenty-nine total, including eight 049s, seventeen 749s, three 1049Ds, and one 1049E.
- United States Air Force Reserve: twenty-six total, including two C-121Gs, nine EC-121Ds, and fifteen EC-121Ts.
- Capitol Airways (Nashville): twenty-two total, including four 749s, five 1049Es, two 1049Gs, and eleven 1049Hs.
- QANTAS Airways: twenty-two total, including seven 749s, four 1049Cs, six 1049Es, three 1049Gs, and two 1049Hs.
- Flying Tigers: twenty-one total, all 1049Hs.
- Capital Airlines (Pennsylvania): nineteen total, including twelve 049s and seven 749s.
- Air India: seventeen total, including seven 749s, two 1049Cs, three 1049Es, and five 1049Gs.
- West Virginia Air National Guard: seventeen total, including nine C-121Cs and eight C-121Gs.
- Panair do Brasil: sixteen total, all 049s.

- Lufthansa: fourteen total, including eight 1049Gs, two 1049Hs, and four 1649As.
- Trans-Canada Airlines: fourteen total, including five 1049Cs, three 1049Es, four 1049Gs, and two 1049Hs.
- Trans International Airlines: fourteen total, including one 049, one 749, three 1049Gs, and nine 1049Hs.
- LAV: thirteen total, including two 049s, two 749s, two 1049Es, and seven 1049Gs.
- Standard Airways: thirteen total, including three 049s, five 749s, one 1049, three 1049Gs, and one 1049H.
- Wyoming Air National Guard: twelve total, all C-121Gs.
- Iberia: eleven total, including three 1049Es and eight 1049Gs.
- Pacific Northern Airlines: eleven total, all 749s.
- Seaboard & Western: eleven total, including four 1049Ds, one 1049E, one 1049G, and five 1049Hs.
- World Airways: eleven total, including seven 1049Hs and four 1649As.
- Avianca: ten total, including six 749s, three 1049Es, and one 1049G.
- Delta Airlines: ten total, including four 049s and six 649s.
- New Jersey Air National Guard: ten total, including nine C-121Cs and one C-121G.
- Slick Airways: ten total, all 1049Hs.
- Varig: ten total, including six 1049Gs and four 1049Hs.

Constellation Registrations

The following table has been compiled to enable the reader to identify any civil-registered Lockheed Constellation from a photograph if the registration can be determined. Every known registration ever applied to a Connie is listed here. In some cases, a single airplane has carried as many as six different registrations over its career.

The first column gives the registration number, which for easy location has been arranged alphabetically by national identification letter, as *AP* for Pakistan, *B* for China, *CC* for Chile, and so on. The US registrations, which differ notably from the international standard, are arranged first by whole numbers, then by numbers followed by letter suffixes. The suffixed registrations are grouped by the suffix letters in the style preferred by members of the American Aviation Historical Association (AAHS) instead of by the system used by the Federal Aviation Administration, N1234, 1234A, 1234B, and so forth.

In column one, the first-assigned registration appears alone; the second and subsequent registrations are identified by parenthetical numbers (2), (3), and so on. If the registration was re-used, the first use is indicated by (#1); second and subsequent re-use of the registration are consecutively numbered. The registrations are followed in column two by the Lockheed model number. Some Connies were modified and were operated under revised model numbers by subsequent owners. It is the model number appropriate to the particular registration that appears in column two. If the airplane was built as a military model that later went on the civil register, the military designation appears in this column.

Column three gives the manufacturer's serial number of the airplane, called the construction number or c/n, which remains constant throughout the life of the airplane regardless of changes in registration or a revised model number. Column four identifies the initial purchaser of the airplane when matching the initial registration. If Lockheed was the first registered owner, as was sometimes the case, the initial purchaser is then identified in parentheses. Other pertinent information, such as the original model number of a redesignated Connie, also appears in column four. Some long airline names necessarily abbreviated in this column are spelled out at the end of the table.

Examples:
The line **N86512/86517 049-46-26 2039/2044 TWA (6)** means that Connies N86512 through N86517 were Model 049-46-26s with serial numbers 2039 through 2044, and that TWA was the initial purchaser of all six.

The line **CF-PXX (3) 1049G-82 4580 From 1049E-55-01** means that CF-PXX is the third registration used by airplane serial number 4580, which at that time was a 1049G-82 that had originally been a 1049E-55-01.

Registration	Model	C/N	Remarks
Pakistan (AP)			
AP-AFQ/RFS	1049C-55	4520/4522	Pakistan Gov't (3)
AP-AJY, RSZ	1049H	4835, 4836	PIA; to Indonesia AF
China (Taiwan, B)			
B-1809 (2)	1049H	4853	
Chile (CC)			
CC-CCA (5)	049-46-59	2069	
Canada (CF, C)*			
CF-AEN (2)	1049H	4821	
CF-BDB (2)	1049H	4809	To C-FBDB 1975*
CF-BFN (2)	1049H	4825	
CF-NAJ, NAK (2)	1049H	4828, 4829	
CF-NAL, NAM (2)	1049H	4831, 4832	
CF-PXX (3)	1049G-82	4580	From 1049E-55-01
CF-RNR (3)	1049C-55	4544	
CF-TEV	1049G-82	4641	Trans-Canada
CF-TEV	1049G-82	4643	Trans-Canada
CF-TEW (2)	1049G-82	4682	

Registration	Model	C/N	Remarks
Canada (CF, C)*			
CF-TEX (2)	1049G-82	4683	
CF-TEX/TEZ	1049H	4850, 4851	Trans-Canada
CF-TGA/TGE	1049C-55	4540/4544	Trans-Canada (5)
CF-TGF/TGH	1049E-55	4563/4565	Trans-Canada (3)
CF-WWH (2)	1049H	4820	
C-GXKO (2)	C-121A	2601	Ex 48-609
C-GXKR (2)	C-121A	2604	Ex 48-612
C-GXKS (2)	C-121A	2609	Ex 48-617
Canadian registration modified in 1975 by moving dash one letter to the left.			
Morocco (CN)			
CN-CCM (2)	749-79-22	2515	
CN-CCN, CCO (2)	749-79-46	2675, 2676	
CN-CCP (2)	749A-79-46	2627	
CN-CCR (2)	749-79-22	2512	
Bolivia (CP)			
CP-797 No. 1 (4)	749A-79-32	2548	
CP-797 No. 2 (3)	1049H-82	4807	
CP-797 No. 3 (5)	1049H-82	4801	

Registration	Model	C/N	Remarks
Portugal (CS)			
CS-TLA/TLC	1049G-82	4616/4618	Tap (3)
CS-TLD (2)	1049H-82	4808	
CS-TLE (5)	1049G-82	4677	
CS-TLF (3,5)	1049G-82	4672	
Cuba (CU)			
CU-C-601, 602	1049G-82	4632, 4633	Cubana
CU-C-631	1049G-82	4675	Cubana
CU-P-573	1049E-55	4557	Pasquel, leased to Cubana
CU-T-547 (2)	049-46-26	2036	
CU-T-601 (2)	1049G-82	4632	Changed from CU-C-601
CU-T-602	1049G-82	4633	Changed from CU-C-602
CU-T-631	1049G-82	4675	Changed from CU-C-631
Uruguay (CX)			
CX-BBM (3)	749A-79-33	2641	
CX-BBN (5)	749A-79-33	2661	
CX-BCS (3)	749A-79-33	2640	
CX-BEM (2)	1049H	4818	
CX-BEN	?	?	Model unknown
CX-BHC (4)	749-79-31	2565	
CX-BHD (3)	749-79-82	2548	
CX-BGP (2)	749A-78-52	2668	
Germany (D)			
D-ALAK	1049G-82	4602	Lufthansa
D-ALAN	1649A-98	1040	Lufthansa
D-ALAP	1049G-82	4637	Lufthansa
D-ALEC	1049G-82	4640	Lufthansa
D-ALEM	1049G-82	4603	Lufthansa
D-ALER	1649A-98	1041	Lufthansa
D-ALID	1049G-82	4647	Lufthansa
D-ALIN	1049G-82	4604	Lufthansa
D-ALOF	1049G-82	4642	Lufthansa
D-ALOL	1649A-98	1042	Lufthansa
D-ALOP	1049G-82	4605	Lufthansa
D-ALUB	1649A-98	1034	Lufthansa
Spain (EC)			
EC-AIN/AIP	1049E-55	4550/4552	Iberia (3)
EC-AMP	1049G-8L	4673	Iberia
EC-AMQ	1049G-8L	4676	Iberia
EC-AQL (5)	1049G	4553	From 1049E-55
EC-AQM, AQN (3)	1049G-82	4644, 4645	
EC-ARN (5)	1049G-8L	4678	
EC-BEN (2)	1049G-82	4519	From 1049C-55-81 via E-55-01
Eire (Ireland, EI)			
EI-ACR, ACS	749-79-32	2548, 2549	Aerlinte Eireann
EI-ADA	749-79-32	2554	Aerlinte Eireann
EI-ADD	749-79-32	2555	Aerlinte Eireann
EI-ADE	749-79-32	2566	Aerlinte Eireann
EI-ARS (4)	1049G-82	4672	
Ethiopia (ET)			
ET-T-35	C-121A	2608	48-616 to Ethiopian Airlines
France (F)			
F-BAZA/BAZD	049-46	2072/2075	Air France (4)
F-BAZE/BAZH	749A-79-46	2624/2627	Air France (4)
F-BAZI/BAZK	749-79-22	2513/2515	Air France (3)
F-BAZL	749-79-22	2538	Air France, to 749A-79-46

Registration	Model	C/N	Remarks
France (F)			
F-BAZM/BAZO	749-79-22	2545/2547	Air France (3), to 749A-79-46
F-BAZP	749-79-22	2550	Air France, to 749A-79-46
F-BAZQ	749-79-22	2512	Air France, to 749A-79-46
F-BAZR (4)	749-79-22	2503	
F-BAZS, BAZT	749A-79-46	2628, 2629	Air France
F-BAZU, BAZV (2)	749A-79-46	2525, 2526	Air France, from 749-79-22
F-BAZX, BAZY	749A-79-46	2527, 2528	Air France, from 749-79-22
F-BAZZ	749A-79-46	2674	Air France
F-BBDT	749A-79-46	2675	Air France
F-BBDU, BBDV	749A-79-46	2676, 2677	Air France
F-BGNA/BGNJ	1049C-55-81	4510/4519	Air France (10)
F-BHBA/BHBH	1049G-82-98	4620/4627	Air France (8)
F-BHBI	1049G-82	4634	Air France
F-BGBJ	1049G-82	4639	Air France
F-BHBK	1649A	1011	Air France*
F-BHBL	1649A-98-11	1020	Air France
F-BHBM, BHBN	1649A-98-11	1027, 1028	Air France
F-BHBO/BHBQ	1649A-98-11	1031/1033	Air France (3)
F-BHBR	1649A-98-11	1036	Air France
F-BHBS, BHBT	1649A-98-11	1044, 1045	Air France**
F-BHMI/BHML	1049G-82	4668/4671	Air France (4)
F-BRAD (4)	1049G-82	4519	From C-55-81 via E-55-01
F-BRNH (2)	1049	4513	From C-55-81 via E-55-01
F-ZVMV (5)	749-79-22	2503	French Air Force
England (G)			
G-AHEJ/AHEM	C-69-5/049E-46	1975/1978	42-94554/94557 to BOAC (4)**
G-AHEN (1, 4)	C-69-5/049	1980	42-94559 to BOAC; later as 049D**
G-AKCE (2)	C-69C-1/049E-46	1971	Ex 42-94550
G-ALAK, ALAL (2)	749A-79	2548, 2549	From 749-79-32
G-ALAM, ALAN (2)	749A-79	2554, 2555	From 749-79-32
G-AMUP (2)	049-46-27	2051	
G-AMUR (2)	049E	2065	From 049-46-27
G-ANNT (2)	749A-79-52	2671	
G-ANTF, ANTG (3)	749A-79	2504, 2505	From 749-79-22
G-ANUP (2)	749A-79	2562	From 749-79-31
G-ANUR (2)	749A-79	2565	From 749-79-31
G-ANUV (3)	749A-79	2551	From 749-79-33
G-ANUX, ANUY (3)	749A-79	2556, 2557	From 749-79-33
G-ANUZ (3)	749A-79	2559	From 749-79-33
G-ANVA (3)	749A-79	2564	From 749-79-33
G-ANVB (3)	749A-79	2589	From 749-79-33
G-ANVD (3)	749A-79	2544	From 749-79-33
G-ARHJ (2)	049-46-27	2051	
G-ARHK (5)	049-46-26	2036	
G-ARVP (5)	C-69-1/049	1967	Ex 43-10315
G-ARXE (6)	C-69-1/049	1965	Ex 43-10313
G-ASYF (2)	749A-79-50	2630	
G-ASYS (2)	749A-79-50	2623	As freighter
G-ASYT, ASYU (2)	749A-79-50	2631, 2632	

*Leased to Air Afrique as TU-TBB.
**Last Connies built; no later registrations.

Registration	Model	C/N	Remarks
Switzerland (HB)			
HB-IBI		?	
HB-IBJ		?	
HB-IEA	C-69-1/049	1965	Ex 43-10313
HB-IEB	C-69-1/049	1967	Ex 43-10305
HB-IED	C-69-1/049	1980	Ex 42-94559

Registration	Model	C/N	Remarks
Haiti (HH)			
HH-ABA (2)	749-79-12	2615	
Dominican Republic (HI)			
HI-129 (3)	649-79-12	2523	
HI-140 (2)	649-79-12	2520	
HI-207 (4)	649-79-12	2522	
HI-228 (2)	1049-53-67	4009	
HI-254 (2)	1049H	4823	
HI-260 (3)	049-46-59	2070	
HI-270 (2)	049-46	2085	
HI-328 (2)	C-121A	2607	Ex 48-614
HI-329 (2)	1049C-55-83	4536	Sprayer
HI-332 (4)	649-79-12	2522	
Colombia (HK)			
HK-162, 163	749A-79-74	2663, 2664	Avianca
HK-175/177	1049E-55	4554/4556	Avianca (were 175X/177X)
HK-184	1049G-82	4628	Avianca
HK-650 (4)	749A-79	2544	From 749-79-33
HK-651 (4)	749A-79	2557	From 749-79-33
HK-652 (4)	749A-79	2564	From 749-79-33
HK-653 (2)	749A-79-52	2645	
South Korea (HL)			
HL-102 (5)	749A-79	2551	From 749-79-33
HL-4002 (2)	1049H	4819	
HL-4003 (2)	1649A-98	1037	Freighter
HL-4006 (2)	1049H	4816	
Panama (HP)			
HP-280, 281	1049G-82	4677, 4678	Thai Airways (2)
HP-467 (7)	1049G-82	4678	
HP-475 (3)	1049E-55	4551	
HP-501 (5)	1649A-98-11	1036	
HP-526 (2)	1049H	4815	
Thailand (SVM, HS)			
HS-TCA (#1)	1049G-82	4644	Thai Airways
HS-TCA (#2)	1049G-82	4672	Thai Airways
HS-TCB (#1)	1049G-82	4645	Thai Airways
HS-TCB (#2)	1049G-82	4677	Thai Airways
HS-TCC	1049G-82	4678	Thai Airways
Argentina (LV)			
LV-FTU, FTV	1049H	4846, 4847	Transcontinental SA (Lse.)
LV-GLH (2)	1649A-98	1006	
LV-GLI (2)	1649A-98	1008	
LV-HCD (2)	1649A-98	1023	
LV-HCU (2)	1649A-98	1009	
LV-IGS (4)	749-79-33	2540	
LV-IIC (5)	749A-79-44	2619	
LV-ILN (2)	1649A-98	1030	
LV-ILW (3)	1049D	4166	
LV-IXZ (5)	1049G-82	4580	From 1049G-55-01
LV-JHF (4)	1049H-82	4801	
LV-JIO (5)	1049H-82	4808	
LV-JJO (5)	1049H-82	4807	
LV-PBH (4)	749A-79-44	2619	Freighter
LV-PCQ (2)	1049D	4166	
LV-PJU (3)	1049H-82	4801	
LV-PKW #1 (3)	049-46-59	2069	
LV-PKW #2 (4)	1049H-82	4807	
LV-PZX (3)	749-79-33	2540	Freighter
Luxembourg (LX)			
LX-IOK (4)	749-79-31	2562	
LX-LGX (4)	1649A-98	1042	
LX-LGY (2)	1649A-98-11	1036	
LX-LGZ (4)	1649A-98	1041	

Registration	Model	C/N	Remarks
United States (NC, N)			
N65 (2)	749A-79-52	2648	
N119 (1st civ)	PO-1W/WV-1	2612	Ex Navy 124437 to FAA
N120 (1st civ)	749A-79-43	2613	Ex Navy 124438 to FAA
N121 (2)	749A-79-52	2654	
N964 (5)	1049G-82	4683	
N1102 (2)	1649A	1001	
N1192 (2nd civ)	PO-1W/WV-1	2612	N119 Renumbered for USAF
N1206 (2nd civ)	PO-1W/WV-1	2613	N120 Renumbered for USAF
N1649	1649A	1001	Lockheed test
N1880	1049H	4820	Dollar Associates (Lease)
N1949 (4)	749-79-31	2565	
N4624 (2)	1049G-82	4617	
N4796 (3)	1649A	1036	
N8021 (2)	1049G-82	4673	
N8022 (2)	1049G-82	4676	
N8023 (2)	1049E-55	4551	
N8024, 8025 (4)	1049G-82	4644, 4645	
N8026 (6)	1049G-82	4678	
N8338 (2)	1049G-82	4616	
N9463 (2)	VC-121A	2602	Ex 48-610
N9464 (1st civ)	C-121A	2601	Ex 48-609
N9465 (1st civ)	C-121A	2604	Ex 48-612
N9466 (1st civ)	C-121A	2607	Ex 48-616
N9467 (1st civ)	C-121A	2609	Ex 48-617
N10401 (3)	749A-79-33	2661	
N10403	749A-79-33	2622	
N25600	C-69/049-39-10	1961	1st Constellation
N38936	C-69-1/049	1962	Ex 43-10310, as 049
N45511 (2)	1649A-98	1034	
N45512 (2)	1649A-98	1040	
N45515 (2)	1049H	4843	
N45516 (2)	1049H	4840	
N45517 (2)	1649A-98	1041	
N45520 (2)	1649A-98	1042	
N51517 (3)	1049G-82	4618	
N54212	C-69C-1/049	1971	Ex 42-94550, as 049
N54214	C-69-5/049	1974	Ex 42-94553, as 049
N67900	C-69/049	1961	Ex 43-10309, as 049-39-10
N67952	C-69-1/049	1963	Ex 43-10311, as 049D
N67953	C-69-1/049	1964	Ex 43-10312, as 049
N70000	C-69-5/049	1974	Fiction
N73544	C-121C	4175	Ex 54-156
N86500, 86501	C-69-10	2021, 2022	Ex 42-94560, 94561, as 049
N86502/86509	049-46-25	2023/2030	TWA (8)
N86510, 86511	049-46-26	2034, 2035	TWA
N86512/86517	049-46-26	2039/2044	TWA (6)
N86520	749-79-22	2503	Lockheed; to Aerovias Guest
N86521	649A-79-60	2642	Chicago & Southern
N86522	649A-79-60	2653	Chicago & Southern
N86523, 86524	649A-79-60	2659, 2650	Chicago & Southern
N86525	649A-79-60	2662	Chicago & Southern
N86526	049-46	2084	Lockheed; to KLM
N86527/86530	749-79-22	2525/2528	Pan Am (4)

Registration	Model	C/N	Remarks
United States (NC, N)			
N86531, 86532 (2)	049-46-59	2068, 2069	
N86533 (3)	049-46-59	2071	
N86535	649A-79-60	2673	Chicago & Southern
N86536	C-69-5	1979	Ex 42-94558, as 049
N86682	1049G-82-118	4679	Qantas
N88831/88833	049-46-26	2031/2033	Pan Am (3)
N88836/88838	049-46-26	2036/2038	Pan Am (3)
N88845/88850	049-46-26	2045/2050	Pan Am (6)
N88855/88862	049-46-26	2055/2062	Pan Am (8)
N88865	049-46-26	2066	Pan Am
N88868	049-46-26	2067	Pan Am
N90607 (2)	749-79-33	2551	
N90608 (2)	749-79-33	2564	
N90621 (2)	749-79-33	2559	
N90622 (2)	749-79-33	2544	
N90623 (2)	749-79-33	2556	
N90624 (2)	749-79-33	2589	
N90625 (2)	749-79-33	2557	
N90814/90826	049-46	2076/2088	TWA (13)
N90827	C-69-1	1965	Ex 43-10313, as 049
N90828/90831	C-69-1	1967/1970	Ex 43-10315, 10316, as 049
N90921/90924	049-46-27	2051/2054	American Overseas (4)
N90925/90927	049-46-27	2063/2065	American Overseas (3)
N91201/91212	749-79-22	2577/2588	TWA (12)
N93164	1049G-82	4581	Freighter, from 1049E-55-01
United States (A, AS, and AV)			
N101A/107A	649-79-12	2518/2524	Eastern (7)
N108A/110A	649-79-12	2529/2531	Eastern (3)
N112A/114A	649-79-12	2533/2535	Eastern (3)
N115A, 116A	749-79-12	2610, 2611	Eastern (2)
N117A/121A	749-79-12	2614/2618	Eastern (5)
N2717A (2)	749A-79-46	2513	From 749-79-22
N2727A (2)	C-69-5	1976	Ex 42-94555, as 049E-36
N2735A (2)	C-69-5	1978	Ex 42-94557, as 049E-46
N2736A (2)	C-69-5	1977	Ex 42-94556, as 049E-46
N2738A	049-46-27	2051	American Overseas
N2739A	049-46-27	2065	American Overseas
N2740A (2)	C-69-5	1975	Ex 42-94554, as 049E-46
N2741A (3)	C-69C-1	1971	Ex 42-94550, as 049E-46
N4192A (3)	1049G-82	4581	From 1049E-55-01
N5595A (3)	749A-79-44	2620	
N5596A (3)	749A-79-44	2619	As freighter
N608AS	C-121B	2600	Ex 48-608, as sprayer
N611AS	C-121A	2603	Ex 48-611, as sprayer
N179AV (3)	1649A-98	1040	
United States (B, C, and CS)			
N2520B, 2521B (3)	049-46	2081, 2082	Ex 90819, 90820, YV-C-AME, AMI
N468C, 469C (2)	1049H	4846, 4847	
N661C (2)	1049G-82	4602	
N1005C (2)	1049E-55	4557	
N1006C	1049H-82	4802	Seaboard & Western

Registration	Model	C/N	Remarks
United States (B, C, and CS)			
N1007C/1010C	1049H-82	4805/4808	Seaboard & Western (4)
N4900C (2)	749A-79-74	2663	
N4901C (3)	749A-79-52	2671	
N4902C (3)	749A-79	2566	From 749-79-32
N4903C	1049G-82	4619	Howard Hughes
N6000C (2)	049-46-59	2020	
N6001C/6005C	749A-79-52	2633/2637	TWA (15)
N6006C	749A-79-52	2639	TWA
N6007C/6015C	749A-79-52	2643/2651	TWA (9)
N6016C/6020C	749A-79-52	2654/2658	TWA (5)
N6021C/6026C	749A-79-52	2667/2672	TWA (6) 6025C to Howard Hughes
N6201C/6214C	1049-53-67	4001/4014	Eastern (14)
N6215C/6218C	1049C-55-83	4523/4526	Eastern (4)
N6219C/6230C	1049C-55-83	4527/4538	Eastern (12)
N6501C/6504C	1049D	4163/4166	Seaboard & Western (4)
N6695C (4)	749A-79-82	2671	
N6696C (2)	749A-79-52	4619	
N6901C/6910C	1049-54-80	4015/4024	TWA
N6911C	1049H	4804	Flying Tigers
N6912C/6915C	1049H	4809/4812	Flying Tigers (4)
N6916C/6918C	1049H	4814/4816	Flying Tigers (3)
N6919C	1049H	4819	Flying Tigers
N6920C	1049H	4822	Flying Tigers
N6921C	1049H	4817	Air Finance Corp.
N6922C	1049H	4825	Air Finance Corp.
N6923C	1049H	4826	California Eastern
N6924C, 6925C	1049H	4852, 4853	Flying Tigers
N6931C	1049H	4813	California Eastern
N6932C	1049H	4823	California Eastern
N6933C	1049H	4826	California Eastern
N6935C, 6936C	1049H	4848, 4849	Slick Airways
N6937C	1049H	4830	Slick Airways
N7023C (3)	1049H	4818	
N7101C/N7120C	1049G	4582/4601	TWA (20)
N7121C/7125C	1049G-82	4648/4652	TWA (5)
N7301C/7309C	1649A-98	1002/1010	TWA (9)
N7310C/7317C	1649A-98	1012/1019	TWA (8)
N7318C/7322C	1649A-98	1021/1025	TWA (5)
N7323C, 7324C	1649A-98	1029, 1030	TWA
N7325C	1649A-98	1035	TWA
N7772C (3)	1049G	4682	
N7776C (2)	1049H-82	4801	
N7777C (2)	1049H-82-133	4803	
N9714C (2)	1049-55-01	4580	
N9716C, 9717C (2)	1049E-55-01	4545, 4546	From 1049C-55-81
N9718C (2)	1049C-55-01	4549	From 55-81
N9719C (2)	1049E-55-01	4574	
N9720C (2)	1049G	4578	From 1049E
N9721C (2)	1049G	4607	
N9722C (2)	1049G-82-118	4679	
N9723C (2)	1049G-82-118	4680	
N9751C (3)	1049G-82	4607	
N9752C (2)	1049H	4850	
N22CS (5)	749A-79	2556	From 749-79-33
United States (D, E, and F)			
N833D (5)	1049G-82	4677	
N563E, 564E (2)	1049H	4833, 4834	
N565E, 566E (2)	1049H	4837, 4838	
N9970E	VC-121A	2602	Ex 48-610
N9812F (3)	749A-79	2559	From 749-79-33
N9813F (4)	749A-79	2589	From 749-79-33
N9816F (4)	749A	2504	From 749-79-22
N9830F	749A-79	2551	From 749-79-33
United States (G)			
N4715G (2)	1049C-55	4543	

Registration	Model	C/N	Remarks
United States (G)			
N6231G	1049G-03-142	4653	Eastern
N6232G	1049G-03-142	4655	Eastern
N6233G	1049G-03-142	4657	Eastern
N6234G/6240G	1049G-03-142	4659/4665	Eastern (7)
N7777G (4)	749A-79	2553	Freighter, from 749-79
United States (H, LM)			
N864H (3)	049D-46-59	2068	From 049-46-59
N1927H	1049H	4821	Air Finance
N8081H	1649A-98	1026	TWA; to cargo
N8082H/8084H	1649A-78	1037/1039	TWA; all to cargo (3)
N9409H (2)	049-46	2074	
N9410H (2)	049-46	2073	
N9412H (2)	049-46	2072	
N9414H (2)	049-46	2075	
N442LM (5)	1049G-82	4581	Freighter
United States (NA, R, and S)			
N101R	1049H	4818	Resort Airlines
N102R	1049H	4824	Resort Airlines
N273R (2)	749A-79-52	2650	
N974R (4)	1649A-98	1040	Freighter
N8742R (2)	1049C-55	4544	
N189S (3)	1049C-55	4541	
N11SR (6)	1049-55-01	4581	Freighter
United States (V Through Z)			
N1552V (4)	749A-79-22	2505	From 749
N1554V (3)	749A-79-32	2555	From 749
N1953V (4)	749A-79-33	2556	From 749
N5172V	1049G-82-102	4572	Northwest
N5173V	1049E-55-01	4573	QANTAS; to 1049G
N5174V, 5175V	1049G-82-102	4576, 4577	Northwest
N5401V	1049H	4839	TWA
N5402V	1049H	4842	TWA
N5403V, 5404V	1049H	4844, 4845	TWA
N173W (3)	1049H	4674	From 1049G-82
N174W (3)	1049H	4636	From 1049G-82
N137X, XR (2, 3)	1049G-82	4650	Note X/XR change
N173X, XR (5, 7)	749A-79-33	2553	From 749; as freighter
N9639Z (2)	1049E-55	4565	
N9640Z (2)	1049G-82	4641	
N9641Z (2)	1049G-82	4643	
N9642Z (3)	1049G-82	4683	
N9733Z	649A-79-60	2573	Chicago & Southern
N9739Z (2)	1049E	4541	From 1049C-55
N9740Z (2)	1049H	4851	
N9741Z (2)	1049C-55	4542	
N9742Z (2)	1049G	4563	From 1049E-55
N9764Z (2)	1049H-82	4674	From 1049G-82
United States (Connies Used by NASA)			
NASA 20	C-121G	4143	54-4065 to NASA
NASA 21	C-121G	4159	54-4076 to NASA
NASA 420	C-121G	4143	54-4065 to NASA
NASA 421	C-121G	4159	54-4076 to NASA
NASA 422	C-121A	2605	48-613 to NASA
Peru (OB)			
OB-R-732 (3)	749-79-12	2614	
OB-R-733 (3)	749A-79	2518	From 649-79-12
OB-R-740 (3)	749A-79	2534	From 649-79-12
OB-R-741 (3)	1049E-55-01	4574	
OB-R-742 (5)	1049C-55-01	4546	
OB-R-743 (2)	749A-79	2535	From 649-79-12
OB-R-771 (2)	749A-79	2521	From 649-79-12

Registration	Model	C/N	Remarks
Peru (OB)			
OB-R-785 (2)	749A-79	2519	From 649-79-12
OB-R-802 (4)	749A-79	2566	From 749-79-32
OB-R-819 (2)	749A-79	2523	From 649-79-12
OB-R-833 (2)	749-79-12	2610	
OB-R-849 (2)	749A-79	2522	From 649-79-12
OB-R-898 (3)	749A-79-46	2627	
OB-R-899 (#1) (4)	749A-79	2549	From 749-79-32
OB-R-899 (#2) (5)	749A-79	2548	From 749-79-32
OB-R-914/917	?	?	Models unknown
OB-WAA-732 (2)	749-79-12	2614	
OB-WAB-733 (2)	749A-79	2518	From 649-79-12
OB-WAC-740 (2)	749A-79	2534	From 649-79-12
OB-LJM-682	?	?	Model unknown
Austria (OE)			
OE-IFA (3)	C-69-1/049	1969	Ex 43-10317
OE-IFE (6)	749A-79	2551	From 749-79-33
OE-IFO (3)	749A-79	2562	From 749-79-31
Netherlands (PH)			
PH-LDD, LDE (2)	749A-79-33	2640, 2641	Re-registered from TFD, TFE
PH-LDG (2)	749A-79-33	2661	Re-registered from LDG
PH-LDK (2)	749A-79-33	2590	Re-registered from TDK
PH-LDN, LDO (2)	749A-79-33	2621, 2622	Re-registered from TDN, TDO
PH-LDP (2)	749A-79-33	2638	Re-registered from TDP
PH-LDR (2)	749-79-33	2540	Re-registered from TEP
PH-LDS, LDT (2)	749A-79-33	2552, 2553	Re-registered from TES, TET
PH-LKA (2)	1049E-55	4553	Re-registered from TFZ
PH-LKB/LKD (2)	1049E-55	4558/4560	Re-registered from TGK/TGM
PH-LKE/LKG	1049G-82	4629/4631	KLM (3)
PH-LKH	1049G-82	4635	KLM
PH-LKI (2)	1049G-82	4644	
PH-LKK (2)	1049G-82	4645	
PH-LKL, LKM	1049H	4840, 4841	KLM (2)
PH-LKN	1049H	4843	KLM
PH-LKP (2)	1049C-55	4501	Re-registered from TFP
PH-LKR/LKY (2)	1049C-55	4502/4509	Re-registered from TFR/TFY
PH-TAU/TAX	049-46-59	2068/2071	KLM (4)
PH-TDA (2)	049-46-59	2071	Re-registered from TAX
PH-TDB	749-79-33	2544	KLM; to 749A-79
PH-TDC	749-79-33	2551	KLM; to 749A-79
PH-TDD/TDG	749-79-33	2556/2559	KLM; (4) to 749A-79
PH-TDH	749-79-33	2564	KLM; to 749A-79
PH-TDI	749-79-33	2589	KLM; to 749A-79
PH-TDK	749-79-33	2590	KLM; to LDK
PH-TDN, TDO	749A-79-33	2621, 2622	KLM; (2) to LDN, LDO
PH-TEN, TEO (2)	049-46	2083, 2084	
PH-TEP	749-79-33	2540	KLM; to LDR
PH-TER	?	2541	Models unknown
PH-TEX	?	2552	Models unknown
PH-TET	749-79-33	2553	KNILM; to LDT
PH-TFD, TFE	749A-79-33	2640, 2641	KLM; (2) to LDD, LDE
PH-TFF	749A-79-33	2652	KLM
PH-TFG	749A-79-33	2661	KLM; to LDG
PH-TFP	1049C-55	4501	KLM; to LKP

139

Netherlands (PH)

Registration	Model	C/N	Remarks
PH-TFR/TFY	1049C-55	4502/4509	KLM; (8) to LKR/LKY
PH-TFZ	1049E-55	4553	KLM; to LKA
PH-TGK/TGM	1049E-55	4558/4560	KLM; (3) to LKB/LKD

Indonesia (PK)

Registration	Model	C/N	Remarks
PK-ALA	?	?	Model unknown

Australia (VH)

Registration	Model	C/N	Remarks
VH-EAA (#1)	749-79-31	2562	Qantas
VH-EAA (#2)	1049E-55-01	4580	Qantas; to 1049G-82
VH-EAB (#1)	749-79-31	2565	Qantas
VH-EAB (#2)	1049E-55-01	4581	Qantas
VH-EAC (#1)	749-79-31	2572	Qantas
VH-EAC (#2)	1049G-82	4606	Qantas
VH-EAD (#1)	749-79-31	2573	Qantas
VH-EAD (#2)	1049G-82	4607	Qantas
VH-EAE (#1) (2)	749A-79	2505	Qantas; from 749-79-22
VH-EAE (#2)	1049E-55-01	4574	Qantas
VH-EAF (#1) (2)	749A-79	2504	Qantas; from 749-79-22
VH-EAF (#2)	1049-55-01	4579	Qantas; to 1049G
VH-EAG	1049C-55-81	4539	Qantas
VH-EAH, EAI	1049C-55-81	4545, 4546	Qantas; (2) to 1049E-55-01
VH-EAJ	1049C-55-81	4549	Qantas; to 1049C-55-01
VH-EAK, EAL	1049E-55-01	4573, 4574	Qantas; EAK to 1049G
VH-EAM	1049H-82	4801	Qantas
VH-EAN	1049H-82-133	4803	Qantas
VH-EAO, EAP	1049G-82-118	4679, 4680	Qantas
VH-EAQ/EAU	1049G-82	4681/4685	Qantas (5)

Rhodesia (VP)

Registration	Model	C/N	Remarks
VP-WAW (3)	1049G-82	4685	

India (VT)

Registration	Model	C/N	Remarks
VT-CQP	749-79-22	2506	Air India; #1 of 3
VT-CQR	749-79-22	2505	Air India; #2 of 3
VT-CQS	749-79-22	2504	Air India; #3 of 3
VT-DAR, DAS	749A-79-44	2619, 2620	Air India
VT-DEU, DEP	749A-79-73	2665, 2666	Air India
VT-DGL, DGM	1049C-55	4547, 4548	Air India
VT-DHL/DHN	1049E-55	4613/4615	Air India; (3) to 1049G
VT-DIL	1049G-82	4646	Air India
VT-DIM, DIN	1049G-82	4666	Air India
VT-DJW, DJX	1049G-82	4686, 4687	Air India; to freighters

Mexico (XA)

Registration	Model	C/N	Remarks
XA-GOQ (2)	749-79-22	2503	Aerovias Guest
XA-GOS	?	?	Model unknown
XA-LIO, LIP (2)	749-79-31	2572, 2573	
XA-MEU (2)	749A-79-44	2620	
XA-MEV (2)	749A-79-73	2665	
XA-MEW (2)	749A-79-44	2619	
XA-NAC (2)	1049G-82	4672	
XA-NAD (2)	1049G-82	4677	
XA-NAF (2)	1049G-82	4678	

Venezuela (YV)

Registration	Model	C/N	Remarks
YV-C-AMA	749-79-34	2560	LAV
YV-C-AME #1 (2)	049-46	2081	LAV
YV-C-AME #2	1049G 82	4636	LAV
YV-C-AMI #1	049-46	2082	LAV
YV-C-AMI #2	1049G-82	4674	LAV
YV-C-AMR	1049E-55	4561	LAV, to YV-C-AMS
YV-C-AMS	1049E-55	4561	LAV; was YV-C-AMR
YV-C-AMT	1049E-55	4562	LAV; to YV-C-ANF
YV-C-AMU	749-79-34	2561	LAV
YV-C-ANB (2)	1049G-82-102	4572	
YV-C-ANC/ANE (2)	1049G-82-102	4575/4577	
YV-C-ANF	1049E-55	4562	LAV; was YV-C-AMT
YV-C-LBI (4)	1049H-82	4808	
YV-C-LBP (2)	1049H-82	4807	

Paraguay (ZP)

Registration	Model	C/N	Remarks
ZP-CAS	?	?	Model unknown

South Africa (ZS)

Registration	Model	C/N	Remarks
ZS-DBR	749A-79-50	2623	South African
ZS-DBS/DBV	749A-79-50	2630/2632	South African (3)
ZS-DTM (3)	1649A-98	1041	
ZS-DVJ (3)	1649A-98	1042	
ZS-FAA (4)	1049G-82	4685	
ZS-FAB (5)	1649A-98	1041	

Ceylon (4R)

Registration	Model	C/N	Remarks
4R-ACH (3)	1049G	4553	From 1049E-55

Israel (4X)

Registration	Model	C/N	Remarks
4X-AKA (5)	C-69-1/049	1965	Ex 43-10313
4X-AKB, AKC (4)	C-69-1/049	1967, 1968	Ex 43-10315, 10316
4X-AKD (3)	C-69-5/049	1980	Ex 42-94557
4X-AKE (3)	049-46-26	2061	
4X-AKF (3)	049-46-26	2036	
4X-AOK (6)	049-46-26	2036	

Cyprus (5B)

Registration	Model	C/N	Remarks
5B-CAJ (6)	1049G-82	4683	

Nigeria (5N)

Registration	Model	C/N	Remarks
5N-07G (2)	1049G-82	4514	From 1049C-55-81
5N-83H (3)	1049G-82	4616	
5N-86H (3)	749A-79-52	2668	

Mauritania (5T)

Registration	Model	C/N	Remarks
5T-TAC #1 (5)	1049G-82	4645	
5T-TAC #2 (4)	1049D	4166	
5T-TAF (2)	1049G-82	4618	
5T-TAG (2)	1049G-82	4642	
5T-TAH (2)	1049G-82	4647	
5T-TAK (2)	1049G-82	4640	

Kenya (5Y)

Registration	Model	C/N	Remarks
5Y-ABF (4)	C-69-5/049	1977	

Senegal (6V)

Registration	Model	C/N	Remarks
6V-AAR (2)	749A-79-46	2538	From 749-79-22

Ghana (9G)

Registration	Model	C/N	Remarks
9G-28 (4)	1049G-82	4683	

ABBREVIATIONS

FAA	Federal Aviation Administration
KLM	Koninklijke Luchtvaart Maatschappi
LAV	Lineas Aeropostal Venezolana
NASA	National Air and Space Administration
PIA	Pakistan International Airlines
QANTAS	Queensland And Northern Territories Aerial Services
TAP	Transportes Aereos Portugueses
TWA	Trans-World Airlines

US Army/Air Force Constellation Serial Numbers

Serial No.	Qty.	Model	C/N	Remarks
42-94549/94559	11	C-C9-5	1970/1980	42-94550 to C-69C
42-94560, 94561	2	C-C9-5	2021, 2022	
43-10309	1	C-69	1961	1st Constellation; to XC-69E
43-10310/10317	8	C-C9-1	1962/1969	43-10310 is plain C-69
48-608	1	VC-121B	2600	
48-609/617	9	C-121A	2601/2609	48-613 to NASA No. 422
51-3836/3845	10	RC-121C	4112/4121	To EC-121C in 1962
52-3411/3425	15	RC-121D	4329/4343	To EC-121D in 1962
53-533/556	24	RC-121D	4348/4371	To EC-121D in 1962
53-3398/3403	6	RC-121D	4372/4377	To EC-121D in 1962

Serial No.	Qty.	Model	C/N	Remarks
53-7885	1	VC-121E	4151	Ex-Navy R7V-1 131650
53-8157, 8158	2	YC-121F	4161, 4162	Ex-Navy RTV-2
54-151/183	33	C-121C	4170/4202	
54-2304/2308	5	RC-121D	4386/4390	To EC-121D in 1962
55-118/139	22	RC-121D	4391/4412	To EC-121D in 1962
54-4048/4079	32	C-121G	*	Ex-Navy R7V-1
67-21471, 21472	2	EC-121R	4382, 4385	Ex-Navy EC-121P (WV-3)
67-21473/21500	28	EC-121R	*	Ex-Navy EC-121K (WV-2)

*Navy airplanes transferred to USAF with random Navy serial numbers and corresponding c/ns instead of in a solid block.

US Navy Constellation Serial Numbers

Navy Serial No.	Qty.	Model	C/N	Remarks
124437, 124438	2	PO-1W	2612, 2613	To WV-1 in 1952
126512, 126513	2	WV-2	4301, 4302	126512 to WV-2E, EC-121L: 126513 to EC-121K in 1962
128323/128326	4	WV-2	4303/4306	To EC-121K in 1962
128434/128444	11	R7V-1	4101/4111 ⎫	
131387/131392	6	WV-2	4307/4312 ⎬ 32 to USAF as C-121G;
131621/131629	9	R7V-1	4122/4130 ⎪ remainder to C-121J in 1962
131632/131659	28	R7V-1	4133/4160 ⎭	

Navy Serial No.	Qty.	Model	C/N	Remarks
135746/135761	16	WV-2	4313/4328	To EC-121K in 1962
137887/137890	4	WV-2	4344/4347	To EC-121K in 1962
137891/137898	8	WV-3	4378/4385	To WC-121N in 1962
140311/140313	3	R7V-1	4167/4169	See R7V-1 above
141289/141333	45	WV-2	4413/4457	To EC-121K in 1962
143184/143225	42	WV-2	4458/4499	To EC-121K in 1962
143226/143230	5	WV-2	5500/5504	To EC-121K in 1962

Bibliography

Andrade, John M. *U.S. Military Aircraft Designations and Serial Numbers.* East Shilton, Leicester, England: Midland Counties Publications, 1979.

Brooks, Peter W. *The World's Airliners.* London: Putnam, 1962.

Clearley, George W., Jr. *Capital Airlines: World's No. 1 Prop-Jet Airline.* 1988.

Davies, R.E.G. *Airlines of Latin America.* Smithsonian Institution Press, 1984.

Airlines of the United States since 1914. Smithsonian Institution Press, 1972.

Pan Am: An Airline and Its Aircraft. New York: Orion Books, 1987.

Fay, Gerald W. *From a -3 to a -10.* Sunflower University Press, 1988.

Federal Aviation Administration. *Specifications for Approved Type Certificates A-763, 6A-5, and 4A-17.* U.S. Gov't Printing Office, Washington, D.C.

Francillon, Rene. *Lockheed Aircraft since 1913.* Putnam and Naval Institute Press, 1982, 1989.

Ginter, Steve. *Naval Fighters No. 8, Lockheed C-121 Constellation.* 1983.

Hardy, M. J. *The Lockheed Constellation.* ARCO Publishing Co., Inc.

Marson, Peter. *The Lockheed Constellation Series.* Air-Britain Publications, 1982.

Morgan, Terry. *The Lockheed Corporation.* ARCO Publishing Co., Inc., 1967.

Serling, Robert. *Eagle: The Story of American Airlines.* St. Martin's Press, 1985.

Shives, Bob, and Thompson, Bill. *Airlines of North America.* Crestline Publishing, 1984.

Swanborough, F. G., and Bowers, P. M. *U.S. Military Aircraft since 1908.* London: Putnam, 1963, 1972, 1989.

U.S. Naval Aircraft since 1911. London: Putnam, 1968, 1976, 1990.

Wixey, Kenneth E. *Classic Civil Aircraft: 1, Lockheed Constellation.* Ian Allen, Ltd., 1987.

Woods, John and Maureen. *Constellation Production List.* Hounslow, Middlesex, England: Airline Publications, 1980.

Yenne, Bill. *Lockheed.* Crescent Books, 1987.

Index